Mini Farming: The Bible for Self-Sufficient, Sustainable Crops for Your Backyard

By Craig W. Van Sickle

Table of Contents

Mini Farming: The Bible for Self-Sufficient, Sustainable Crops for Your Backyard...1

By Craig W. Van Sickle...1

Mini Farming: The Bible for Self-Sufficient, Sustainable Crops for Your Backyard...2

  Book Description...2

  Overview of Intensive Agriculture...2

  How Mini Farming Works for You...12

  Raised Beds...19

  Soil Composition and Maintenance...21

  Plant Nutrients...24

  Yield...27

  Watering and Irrigation...28

  Pest and Disease Control...34

  Seed Starting...35

Mini Farming: The Bible for Self-Sufficient, Sustainable Crops for Your Backyard

Book Description

The basics you need, from composting to harvest, to produce more than 80% of what an average family needs for self-sufficiency using sustainable, organic farming methods.

"Mini Farming: The Bible for Self-Sufficiency, Sustainable Crops For Your Backyard' guides you to a holistic approach to growing garden crops for your family – while earning additional income from the sale of the excess production.

Whether a beginner or an experienced in mini farming, this book covers everything you need to know to get started:

· Soil Basics
· Crop planning and location relative to your house
· Seeds
· Composting
· Incorporating Bees
· Small Animals (Should You Raise…?)
· Crop Rotation
· Selling Your Produce and much more

Overview of Intensive Agriculture

Subsistence agriculture occurs when farmers grow food crops to meet their families' needs on smallholdings. Subsistence agriculturalists target farm output for survival and mostly local requirements, with little or no surplus.

Intensive farming is a technique to yield high productivity by keeping large numbers of livestock indoors. It is an agricultural system aiming to get maximum yield from the available land. This farming technique is also applied in supplying livestock.

Intensive agriculture, in agricultural economics, is a system of cultivation using large amounts of labor and capital relative to land area. Large amounts of labor and capital are necessary for applying fertilizer, insecticides, fungicides, and herbicides to growing crops. Capital is significant to acquiring and maintaining high-efficiency machinery for planting, cultivating, and harvesting, as well as irrigation equipment where required.

Optimal use of these materials and machines produces significantly greater crop yields per unit of land than extensive agriculture, which uses little capital or labor. As a result, a farm using intensive agriculture will require less land than an extensive agriculture farm to produce a similar yield. In practice, however, intensive agriculture's increased economies and efficiencies often encourage farm operators to work vast tracts to keep their capital investments in machinery productively engaged.

On the level of theory, the increased productivity of intensive agriculture enables the farmer to use smaller land areas close to the market, where land values are high relative to labor and capital. This is true in many parts of the world. Suppose costs of labor and capital outlay for machinery and chemicals, and costs of storage (where desired or needed) and transportation to market are too high. Farmers may find it more profitable to turn to extensive agriculture in that case.

However, in practice, many small-scale farmers employ some combination of intensive and extensive agriculture, and many operate relatively close to markets. Many large-scale farm operators, especially in relatively vast and agriculturally advanced nations such as Canada and the United States, practice intensive agriculture in areas where land values are relatively low and at great distances from markets. They farm enormous tracts of land with high yields. However, in such societies, overproduction (beyond market demands) often results in diminished profit due to depressed prices.

Intensive Agriculture

Though intensive agriculture elicits images of vast tracts of land, monoculture, pesticides, and barns filled with thousands of animals hardly able to stand, this is not how the practice started. Intensive farming originated in the ancient civilizations of Egypt, Mesopotamia, India, Pakistan, North China, Mesoamerica, and Western South America with the creation of water management systems and the domestication of large animals that could pull plows. In more recent years, and especially since industrialization, intensive agriculture has also come to be characterized by various other practices such as heavy pesticide use, rotational grazing, and concentrated animal feeding operations.

What Is Intensive Agriculture?

Intensive agriculture is a method of farming that uses large amounts of labor and investment to increase the yield of the land. In an industrialized society, this typically means using pesticides, fertilizers, and other chemicals that boost yield and acquiring and using machinery to aid planting, chemical application, and picking. In theory, this reduces the amount of land needed for an economically viable farm to grow crops or raise animals. However, in countries such as the United States and Canada, these methods are often used to overproduce products as companies attempt to increase their market share. Profit is then diminished so that farmers must continue overproducing to stay economically viable and often seek compensation for low profits via government subsidies.

What Are the Characteristics of Intensive Agriculture?

Pasture Intensification

Pasture intensification is the increase in value and production due to inputs such as money, labor, and pesticides, specifically in the pastures on which farmed animals graze. Historians believe that pasture intensification, as well as agricultural intensification more broadly, was a necessary step in creating our modern societies. New farming methods and increasing yield allowed for larger populations to grow.

The most common and effective way of increasing inputs throughout history has been to plant or graze more land, increasing farm yield. However, increasing the amount of land used can have severe consequences for biodiversity, which is lost when native plants and grasses are cleared to make room for grazing.

In recent years there has been an increased interest in methods of intensification that reduce some of its adverse effects. These include incorporating crops, such as soybeans or other legumes, in pastures in which cows graze.

Rotational Grazing

Rotational grazing is a type of pasture intensification that entails breaking grazing areas into smaller paddocks. Farmed animals are rotated through the different paddocks one by one, allowing those paddocks not in use to recuperate and regrow foliage. This is distinct from traditional grazing, as typically, cattle are allowed to free-graze an entire pasture, which results in selective grazing and does not provide adequate time for regrowth. This leads to more land needed to support the farmed animals.

Concentrated Animal Feeding Operations
Concentrated animal feeding operations are the predominant type of animal farm in industrialized agriculture systems. They consist of large numbers of animals, generally enclosed in feedlot operations. Instead of grazing and gathering their food, the animals have food brought to them. The animals are confined in small spaces with little room or opportunity to express natural behaviors.

Crop Irrigation

Crop irrigation uses artificial systems to control water application and make up for any shortage of natural rainfall. In California, more than nine million acres of land are irrigated, accounting for 80 percent of the water used by businesses and homes. The heavy use of irrigation to grow crops in areas that cannot sustain them naturally creates risks and challenges, primarily because of the ongoing threat of drought in many such places.

Genetically Modified Organism (GMO) Seeds

Many of the most abundant crops in the United States are species that have been genetically modified. In 2018, 94 percent of all soybeans, 94 percent of cotton, and 92 percent of corn planted in the country were genetically modified. Generally, seeds and crops are genetically modified to be larger, more pest-resistant, or tolerate herbicides better.

Use of Agrochemicals

Modern-day industrialized intensive agriculture uses large amounts of pesticides and fertilizers. These chemicals wreak havoc on ecosystems, polluting water and killing essential species such as bees and ladybugs.

Intensive Agriculture Examples

Livestock

Most of the farmed animals in the United States live a significant portion of their lives on industrial factory farms that use various intensive methods to produce more meat, dairy, or eggs for less money. One such method is keeping the animals enclosed in small spaces and delivering their food to them. This forces them to grow more quickly and reduces the need for space. Another example of intensive animal agriculture methods is selectively bred animals that grow more quickly than naturally occurring breeds and get large enough for slaughter in a shorter period.

Aquaculture

Intensive agriculture is apparent in every part of the industry, and aquaculture is no exception. One example is the standard practice of housing extremely high densities of fish in artificial tanks, allowing the farmers to control feed, oxygen levels, and various other factors leading to an increase in yield.

Crops

There are several ways that farmers who grow crops use intensive agriculture to produce higher yields. Tactics include the use of pesticides, insecticides, fertilizers, irrigation, and the use of genetically modified seeds.

Intensive Versus Extensive Agriculture
Methodology

Intensive farming invests considerable resources and labor into small tracts of land to increase yield. On the other hand, extensive agriculture employs larger tracts of land with lower labor and input costs associated with it.

Location

Traditionally, one advantage of intensive agriculture is that because it requires less land, you can produce yield closer to the market than farms using extensive agriculture. However, most modern farms using intensive methods in higher-income countries operate on a large scale, often on thousands of acres and in areas far from where consumers live.

Farm Land Area

In theory, extensive farming requires much more land than intensive agriculture, as additional chemicals, machinery, and labor are not applied to increase yield. However, due to the shift in farming techniques favoring intensive methods for larger tracts of land, both intensive and extensive farming techniques use large amounts of land today.

Inputs

Intensive farming requires more significant inputs than extensive farming. Intensive farms tend to use more labor, agrochemicals, and special seeds or breeds of animals. Extensive farming largely relies on the natural fertility of the land and the natural behaviors of the animals.

Profitability

Modern-day intensive farming has sought to produce massive amounts of food as cheaply as possible. This has resulted in the overproduction of many food items, which drives the market price down. For extensive farming to be profitable large amounts of land are required. Because of this, extensive methods tend to be used most frequently, with low population densities and inexpensive land.

Productivity (Yield/Hectare)

Unsurprisingly, because intensive agriculture aims to maximize land productivity, the yield per hectare is higher than that of farms that utilize extensive methods.

Environmental Impact

Both extensive and intensive farming can have negative environmental impacts. Extensive farming requires large amounts of arable land and has often led to deforestation, while intensive farming involves chemicals that negatively impact the environment and native species. Feed for intensively farmed animals is a growing factor in deforestation as well. Growing the food used to feed animals in intensive farming also leads to deforestation in South America and increasingly in other areas of the world.

Perceived Problems with Intensive Agriculture

Animal Cruelty

Billions of animals in the United States are raised on factory farms that employ intensive methods to increase profitability. Often they are confined in small spaces w here they can barely move. Standard procedures include debeaking, castration, tail docking, and dehorning. These frequently occur without sedation, causing pain for the animals that endure them.

Deforestation

Because intensive agriculture has shifted from focusing on maximizing the productivity of small plots of land to application on farms spanning thousands of acres, it can often drive deforestation even before one considers the sources of animal feed. Because the land must be easily accessible for planting, watering, and fertilizing, you must remove trees to create large expanses of flat land. Growing corn and soybeans is a leading cause of deforestation globally.

Human Health

Exposure to the pesticides that intensive agriculture tends to use in large quantities can have many adverse effects on human health. These include irritation to the skin and eyes and adverse effects on the nervous and endocrine systems. The mismanagement of the large amounts of manure produced by intensive animal agriculture can also lead to health problems in surrounding communities.

Pest and Weed Resistance

Following repeated applications of a particular pesticide or herbicide, many plant and animal pest species can build up resistance to these chemicals. This often results in more potent chemicals being used to destroy the target species or a more significant amount or higher concentration of the chemicals being applied to obtain the desired results.

Soil Degradation

Intensive farming can contribute to soil degradation, as land tends to be planted repeatedly without providing a break for the soil to recover its nutrients. This often results in the increased use of fertilizers to make up for the lack of nutrients in the soil.

Water Pollution

Intensive farming methods can contribute considerably to water pollution. Every year the animals on factory farms in the United States produce billions of gallons of waste. With nowhere else to go, this waste tends to be stored in large cesspools or sprayed over fields. Both disposal systems can result in water pollution, especially when the waste makes its way into rivers, lakes, or other bodies of water.

Climate Change

The increase in intensive farming methods around the world was noted in 2020 as threatening the world's chance of meeting the terms of the Paris agreement. The use of artificial fertilizers and the farming of animals—especially cows, which produce large quantities of methane and are often fed with grains farmed on deforested land—are causes of increasing emissions of greenhouse gases.

Harm to Smaller Farms

The rise of intensive agriculture has dealt a serious blow to small farms. Because the larger corporate enterprises can afford to produce crops and animals on a much larger scale, they can sell them for a lower market price while still making a profit. Smaller farms, however, have been left behind, as they do not tend to produce enough to accept such low prices. This has contributed to many farmers leaving the industry and other social effects on farming communities.

The intensification of farming has played an important role in the history of agriculture. It allowed farmers to feed growing communities around the world. However, intensive agriculture as we know it today is no longer sustainable or necessary. The methods employed can have negative impacts on the environment, human health, animal lives, and communities caused by the heavy use of chemicals, which is the trademark of modern-day intensive farming. We can reduce our support of the industry by purchasing locally grown foods from farms that bolster the local economy and employ more environmentally friendly methods of production.

Intensive Farming: Advantages & Disadvantages

Intensive animal farming is a technique to yield high productivity by keeping large numbers of livestock in feed lots. This farming technique is also applied in supplying livestock. It is also an agricultural system aiming to get maximum yield from the available land. You could say that this technique produces food in large quantities with the help of chemical fertilizers and pesticides that are appropriately used to save such agricultural land from pests and crop diseases.

Advantages of Intensive Farming

1. One of the significant advantages of this farming technique is that the crop yield is high.
2. It helps the farmer efficiently supervise and monitor the land and protect his livestock from being hurt or hounded by dangerous wild animals.
3. With the introduction of intensive farming, farm produce, such as vegetables, fruits, and poultry products, have become less expensive. It also aids in solving the global hunger problems to a great extent. This means that ordinary people can now afford a balanced and nutritious diet.
4. The introduction of intensive farming creates the opportunity to utilize less space, equipment, and other inputs required for farming, making the business of farming more cost efficient.
5. Another advantage is that greater productivity in food production is possible utilizing smaller plots of land.
This leads to economies of scale and directly contributes towards meeting the ever-growing demand for food production.

Disadvantages of Intensive Farming

1. Intensive farming involves various chemical fertilizers, pesticides, and insecticides.
2. It can lead to overcrowding of feed lots if the holding facilities are not managed properly, which can lead to local environmental pollution and potential outbreaks of diseases and infection.

3. The overuse of chemical fertilizers can contaminate the soil and water bodies such as lakes and rivers.
4. If not properly handled, using adequate protection, the use of pesticides and chemical fertilizers can also affect the workers (who spray the pesticides) and the people residing nearby.

How Mini Farming Works for You

The Business of Mini-Farming

Many homeowners undertake the task of gardening or small-scale farming as a hobby to get fresh produce and possibly save money over buying food at the supermarket. Unfortunately, the most common gardening methods are so expensive that even some enthusiastic garden authors' outright state that gardening should be considered, at best, a break-even affair.

Looking at the most common gardening methods, these authors are correct. Standard gardening methods are considerably more expensive than necessary because they were initially designed to benefit from the economies of scale of corporate agribusiness. When home gardeners try to use these methods on a smaller scale, it's a miracle if they break even over a several-year period, and it is more likely they will lose money.

The Economics of a Mini-Farm

The cost of tillers, watering equipment, large quantities of water, transplants, seeds, fertilizers, and insecticides adds up quickly. Balanced against the fact that most home gardeners grow only vegetables, and vegetables make up less than 10 percent of the calories an average person consumes, it quickly becomes apparent that even if the cost of a vegetable garden were zero, the amount of actual money saved in the food bill would be negligible.

For example, if the total economic value of the vegetables collected from the garden in a single season amounted to about $350, even if you could produce the vegetables for free, the economic benefit would amount to only $7 a week when divided over the year. The solution to this problem is to cut costs and increase the value of the end product. Using this combination, the economic equation balances favor of the gardener instead of the garden supply store. It becomes possible to supply all of a family's food except meat (if you eat it) from a relatively small garden. According to the USDA, the average yearly cost to feed a family of three is $8,140.92—on a low-cost plan. Increase that to a liberal plan, and we're looking at an average cost of more than **$12,000 annually**. Understanding that food is purchased with after-tax dollars, it becomes clear that home agricultural methods that take a significant chunk out of that figure can make a difference.

The key to making a garden work for your economic benefit is to approach mini-farming as a business. No, it is not a business in the sense of incorporation and taxes unless some of its production is sold. But think of it as a business in that reducing your food expenditures can have the same net effect on finances as income from a small business. Like any small business, it could earn or lose money depending on how it is managed.

Seeds and Seedlings

Garden centers are flooded every spring with gardeners buying seedlings. For hobbyist gardeners, this may work well because it allows a quick start with minimal planning. But it's a bad idea for the mini-farmer who approaches gardening as a small business.
In my garden this year, if I plan to grow 48 broccoli plants, seedlings from the garden center would cost $18 if discounted, possibly more

than $30. Even the market's most expensive organic broccoli seeds cost less than a dollar for 48 seeds. Growing seedlings to transplant at home drops their effective cost from around $30 down to $1. Adding the cost of soil and containers, the cost is still only about **$2** for 48 broccoli seedlings. Considering that a mini-farm would require transplants for dozens of crops, from onion sets to tomatoes and lettuce, it quickly becomes apparent that growing from seed saves hundreds of dollars a year.

The two primary seed/plant varieties available are hybrid and open-pollinated. Open-pollinated varieties produce seeds that duplicate the plants that produced them. Hybrid plant varieties produce seeds that are, at best, unreliable and sometimes sterile and, therefore, often unusable.

Although hybrids have the disadvantage of not producing viable seed, the advantage that makes them worthwhile is "hybrid vigor," a poorly understood phenomenon in plants where a cross between two varieties can yield far more vigorous and productive offspring than either parent. Using hybridization, seed companies can then deliver varieties incorporating disease resistance into a particularly good-tasting variety. So why not just use hybrid seeds? There is no measurable increase in vigor in hybrids for plants that generally self-pollinate, such as peppers and tomatoes. The hybrids are just a marketing avenue; buying hybrids raises costs and forces you to repurchase seeds next year.

Another reason to save seeds from open-pollinated plant varieties is that each year as you save seeds from the best-performing plants, you will eventually create varieties with genetic characteristics that work best in your particular soil and climate. That's a degree of specialization money can't buy. Of course, there are cases where hybrid seeds outperform open-pollinated varieties. Hybrid seeds that manifest pest- or disease-resistant traits can be a good choice when those pests or diseases cause ongoing problems. When using hybrid seeds eliminates the need for synthetic pesticides, they're a good choice

Garden Intensively

Many intensive gardening methods have been well-documented over the past century. All methods have in common the spacing of growing plants much closer together than traditional row methods. This closer spacing significantly decreases the land required to grow a given quantity of food, reducing requirements for water, fertilizer, and mechanization. Because plants are grown close enough to form a sort of "living mulch," the plants shade out weeds and retain moisture better, thus decreasing the amount of work required to raise the same amount of food. Intensive gardening techniques make a big difference in the space required to provide a person's food. Current agribusiness practices require 30,000 square feet (or 3⁄4 acre) per person. Intensive gardening practices can reduce the amount of space required for the same nutritional content to 700 square feet, plus another 700 square feet for crops explicitly grown for composting. That's only 1,400 square feet per person. Intensive gardening techniques are the key to self-sufficiency on a small lot.

Compost

Because growing so many plants in such little space puts heavy demands on the soil, all intensive agriculture methodologies pay particular attention to maintaining soil fertility. Standard agribusiness practices would suggest buying commercial fertilizers from outside the farm. While there are other highly worthwhile reasons for avoiding the use of nonorganic fertilizers (including human health and environmental damage), economics alone makes a good case for avoiding synthetic fertilizers. A mini-farm with a well-managed soil-fertility plan can drastically reduce the need to purchase fertilizer, thereby reducing one of the most significant costs associated with farming. A certain amount of fertilizer may always be required, especially at the beginning, but using organic fertilizers and creating compost can reduce fertilizer requirements to a bare minimum.

Preserving soil fertility consists of growing crops specifically for compost value, growing crops to fix atmospheric nitrogen into the soil, and composting all crop residues possible (along with the specific compost crops).

Calorie-Dense Plants:

Vegetables provide only about 10 percent of the average American's caloric intake.. Because of this, a formal vegetable garden may supply excellent produce and rich vitamin content, but the economic value of the vegetables won't significantly reduce your food bill. The solution is to grow crops that provide a higher proportion of caloric needs, such as fruits, beans, grains, and root crops, such as potatoes and onions.

Meat:

Most Americans obtain at least a portion of their protein from eggs and meat. Agribusiness meats are often produced using practices and substances (such as growth hormones and antibiotics) that worry many people. Certainly, factory-farmed meat is very high in the least healthy fats compared with free-range, grass-fed animals. The problem with meat, in an economic sense, is that each calorie of meat generally requires two to four calories of feed. This sounds, at first, like an inefficient use of resources, but it isn't as bad as it seems. Most livestock, including poultry, gets a substantial portion of its diet from foraging. Poultry will eat the ticks, fleas, spiders, beetles, and grasshoppers and dispose of the farmer's table scraps. If meat is raised on-premises, the mini-farmer has to raise enough food to make up the difference between feed needs and what's obtained through scraps and foraging.

Fruit:

Many fruits can be grown in most parts of the country: apples, grapes, blackberries, pears, and cherries, to name a few. Dwarf fruit tree varieties often produce substantial fruit in only three years and take up comparatively little space. Grapes native to North America, such as the Concord grape, are hardy throughout the continental United States. Some varieties, such as Muscadine grapes, grow prolifically in the South and offer unique health benefits. Strawberries are easy to grow and attractive to youngsters. You can easily preserve fruits; many can also be stored whole for a few months using a root cellar.

Market Crops:

If you adopt organic growing methods, you can get top dollar for crops delivered to restaurants, food cooperatives, farmer's markets, etc. According to John Jeavon's research described in The Complete Biointensive Mini-Farm, a U.S. mini-farmer could expect to earn **$2,079** in income from the space required to feed one person, in addition to actually feeding the person. Assuming a family of three and correcting for USDA-reported rises in the value of food, that amounts to about **$10,000 a year**, using a six-month growing season. A mini-farm that sets aside only 2,100 square feet for market crops could gross an average of **$11,289 annually**. (It's worth noticing that two authorities arrived at similar numbers for expected income from vegetable sales—about $5 a square foot.)

Extend the Season

Many people don't realize that most of Europe, where greenhouses, cold frames, and other season extenders have been used for generations, lies north of most of the United States. Maine, for example, is at the same latitude as southern France. The difference in climate has to do with ocean currents, not latitude, and latitude is the most significant factor in determining the success of growing protected plants because it determines the amount of sunlight available. Essentially, anything you can do in southern France can be done throughout the continental United States.
Extending the season allows for earlier starts and later endings to the growing season, netting more food. The secret lies in working with nature, not against it. Any attempt to build a super insulated, heated tropical environment suitable for growing bananas in Minnesota in January will be prohibitively expensive. A simple unheated hoop house covered with plastic is relatively inexpensive and will work exceptionally well with crops selected for the climate.

The Economic Equation

This information is based on math presented in Mini Farming by Brett L. Markham, from which this section is adapted, but has been updated with more recent figures. It is only an example, but it shows the economic impact of mini-farming on a household's bottom line. According to the Social Security Administration, as of 2014, the median U.S. nonfarm wage earner makes $28,851. Assuming a roughly 25 percent tax rate, this person takes home around $21,638. Nationwide, childcare costs $300 to $1,564 a month—on average, $11,666 a year. Assuming a school-age child means the average worker has $9,972.46 post-tax income.

Though there are other justifications for adopting mini-farming, it may make economic sense for one member of a working couple to become a mini-farmer if the net economic impact of the mini-farm can replace the income from the job. Mini-farming may not be an excellent economic decision for those in highly paid careers. But mini-farming can have a sufficient net economic impact that most standard occupations can be replaced. Mini-farming is also sufficiently time-efficient that you could use it to remove the need for a second job.

According to Census Bureau statistics from 2014, the average household size in the United States is 2.54 people. Let's round that up to three for ease. According to statistics from the USDA, the cost of feeding a family of three with two adults and one child with a low-cost plan amounts to **$8,140.92 per year** (this cost would likely go up if the family was committed to eating organic). A mini-farm that supplied 85 percent of those needs would produce a yearly economic benefit of roughly **$7,000 annually**.

This means the mini-farmer has produced about 70 percent of the take-home pay she would have earned at a job, in much less time, without commuting and without paying for childcare. If the farm also dedicated 2,100 square feet and five hours a week to market crops, it could earn an additional $10,000 during a standard growing season. The mini-farmer also gains back more than 1,500 hours a year that you can use to improve your quality of life in many ways; gains a much healthier diet; gets regular exercise; and gains a measure of independence from the typical employment system. It's impossible to attach a dollar value to that.

For families who want to have a parent stay at home with a child, mini-farming may make it possible—and make money in the process, by having whichever parent earns the least money from regular employment go into mini-farming. For healthy people on a fixed income, it's a no-brainer.

Raised Beds

Raised beds can be as humble or creative as you like. A raised bed planter can be a permanent fixture for perennial plants to settle in and mature. The initial cost of getting your raised bed set up will depend on how elaborate you make it, but once in place, raised beds are no more expensive to maintain than traditional gardens. They offer many benefits. Raised bed gardens can fit just about any space. With creativity, you can create an entire garden sitting area, complete with a potting shed and lamppost. Add a bench section and you have seating for an outdoor dining area. As the plants fill in and the wood weathers, this garden will have a natural, rustic appearance.

When making a raised bed, instead of going in-ground, you can place it where the sun or shade is the best for the plants you want to cultivate. You can also prevent tunneling pests from decimating your plants. Plants can be healthier and more productive in a raised bed because you can control the soil quality and water drainage. You can even sit and garden if you build the sides wide enough to make a bench. Positioning can make it easier for those with back problems to tend the plants. Raised beds of brick or wood, as pictured, can also enhance the design of your homestead or backyard. Another great advantage of raised bed gardens is that they sit well above the underground frost line, so the soil warms up faster in the spring, and you can start planting sooner. The material used for your beds makes a difference here: metal holds more heat from the sun. But grow bags are a good option as they don't freeze solid, and the soil in them defrosts rather quickly. Also, it is a great way to provide the heat needed to grow Mediterranean plants like sage and lavender. Grow bags may seem too easy, but within minutes you could have a significant raised bed garden!

Spiral gardens are a popular permaculture technique. They increase the amount of usable planting area without taking up more ground space in your garden. You can easily build them out of stone, brick, or wood or simply pile up the soil. The unusual shape and swirl of plants make for an eye-catching focal point in your garden. You can grow anything using a spiral design.

One of the easiest ways to create raised bed gardens is using animal feeding troughs. No assembly is required, but drill some drainage holes in the bottom before adding the soil. The metal gives the garden an industrial look and conducts heat, warming the soil in the spring. You can use new or used troughs depending on availability and your desired look. Depending on what you choose to grow, the plants may need extra water during the hottest part of summer.

Types of raised garden beds

Before shopping for a raised garden bed, you must decide which type is right for you and your outdoor space. Venelin Dimitrov, senior product manager for gardening company Burpee, said there are three types of raised garden beds: raised ground beds, supported raised beds, and containerized raised beds. All three work to lift plants and their root systems above ground.

Raised ground beds are flat-topped mounds of soil 6 to 8 inches high and do not have support frames. Dimitrov said they're the simplest above-ground gardening option since they only require soil and are sometimes referred to as built-in raised beds. He noted that they're specifically useful for gardeners with large areas for planting but do not want the added expense of building support frames.

Supported raised beds consist of a mound of soil surrounded by a supportive edge or frame. Dimitrov said these beds are handy for sloped or uneven ground — you can build a flat surface on top of an uneven surface with this raised bed. Experts state that containerized raised beds are what most people typically think of when discussing raised garden beds. They are essentially large planters or pots. Containerized raised beds have taller sides and a base, and you can use them on lawns, walkways, decks, patios, driveways, and porches. Dimitrov said you should be sure that wherever you put the raised bed can tolerate its weight and the moisture it may give off. He noted that you might need more soil to fill a containerized raised bed, but they're versatile and work well in high-traffic areas.

Soil Composition and Maintenance

Preparing and maintaining the health of your soil is the equivalent of taking your vitamins, exercising, and drinking plenty of water. We take supplemental vitamins to maintain ideal nutrition, exercise allows oxygen to flow through the blood to ensure our tissue is healthy, and the water replenishes our bodies and assists oxygen transport, among other vital functions. Other living organisms, including trees, plants, microbiota, and decomposers, also need nutrients, oxygen, and water to maintain their systems. By preparing your soil at the beginning of the growing season and carrying out a maintenance schedule for the rest of the season, your garden should maintain its health and production.

First, we need to know a little bit about soil. Soil is a mixture of sand, clay, and loam, depending on your geographic and geologic area. Your soil may have a predictable composition. Healthy soil also contains air, water, and a small amount of organic material. Loam soils are considered ideal because they are a good combination of sand and clay. The sand allows air and water to penetrate the earth, while the clay holds nutrients and moisture. You can purchase unique products at your local gardening store to amend your soil if you have clay soil. People have successfully added builder's sand and heavy compost to their clay soils. Builder's sand is different from ocean sand or playground sand. The builder's sand is coarse and large-grained. Adding this to fine clay should disrupt the settling pattern of clay. Also, by adding coarse and heavy compost, you will add nutrients to the soil and provide a biodegradable aggregate, disrupting the settling pattern of clay.

The manner of soil preparation described above is ideal for growing vegetables. Most annual vegetables are expected to germinate, flower, and produce fruit or mature within a short time frame. They can take advantage of the freshly applied organic matter before it begins to decompose further and lose its nutritional value. Their roots spread quickly but not as deeply or extensively as a fruit tree. If you intend to plant vegetables, use the above mixture to achieve optimal results. If you're planting fruit trees, there are two schools of thought related to soil prep. Some say you should add compost and soil inputs to the ground when planting. This could give the tree a head start by having nutrients close to its roots. However, when the roots reach the native soil, they might go into shock. The second method is planting fruit trees directly into the native clay soil. The tree may take some time to acclimate to the parent soil, but its growth should proceed uninterrupted after it becomes accustomed to its surroundings.

Once you have prepared your soil, it's time to develop a maintenance schedule. Soil maintenance for vegetable beds is relatively simple. Compost can be added to the soil before planting and then once more during the growing season, preferably when plants are flowering. If you do not have compost readily available, you can use plant food instead. Keep weeds pulled so they don't consume the nutrients intended for your plants. Adding tree stakes to the tree's base can also maintain fruit trees. About three tree stakes will suffice per tree, but read the instructions on the packaging just in case. You should add tree stakes or tree food about three months before the fruiting season of your particular type of tree.

Soil Fertility: Definition, Types, and Maintenance
Definition of Soil Fertility:

According to Wikipedia, the definition of soil fertility is as follows: "The ability to supply essential plant nutrients and water in adequate amounts and proportions for plant growth and reproduction," and "The absence of toxic substances which may inhibit plant growth."

Types of Soil Fertility:
1) Inherent or Natural Fertility:
a) The nutrients that the soil contains in its natural state are known as inherent fertility.
b) The essential plant nutrients, nitrogen, phosphorus, and potassium are the basic macronutrients needed for crops' normal growth and yield. Inherent fertility can be a limiting factor in plant growth.
2) Acquired Fertility:

a) The fertility developed by applying manures and fertilizers, utilizing tillage, irrigating, improving drainage, etc., is known as acquired fertility.

b) Acquired , especially in the application of fertilizers, has a point of decreasing returns where additional units do not create an increase in crop yields.

Factors Affecting Soil Fertility:

The factors that are affecting soil fertility may be of two types:
a) Natural Factors:
Natural factors influence soil formation, and are related to the proper use of land. The factors affecting the fertility of soil are parent material, climate, and vegetation, topography, the inherent capacity of soil to supply nutrients, the physical condition of the soil, soil age, micro-organisms, availability of plant nutrients, soil composition, organic matter, soil erosion, cropping system and favorable environment for root growth.

(ii) Artificial Factors:
These factors include conservation techniques to improve soil tilth, and drainage methods to decrease moisture levels of water logged soils, among other techniques.

Maintenance of Soil Fertility:
Maintenance of soil fertility is a significant problem for our farmers. The cultivation of a particular crop in the same field year after year decreases soil fertility. To increase soil fertility, it is necessary to check the loss of nutrients and increase the soil's nutrient content. You must adequately follow the following things to increase soil fertility.
(i) Proper use of land,
(ii) Good tillage,
(iii) Crop rotation,
(iv) Control of weeds,
(v) Maintenance of optimum moisture in the soil,
(vi) Control of soil erosion,
(vii) Cultivation of green manure crops,
(viii) Application of manures,

(ix) Cultivation of cover crops,
(x) Removal of excess water (drainage),
(xi) Application of fertilizers,
(xii) Maintenance of proper soil reaction.

Soil contains air, water, minerals, and plant and animal matter, both living and dead. Your soil might vary from one part of your land to another. Ideally, a "good horticultural soil" contains 50% solid material.

One of the keys to maintaining native and climate-appropriate plants in your sustainable landscape is soil health. The following are some of the steps you can take to increase the fertility of the soil:

Improve the humus of the soil by regularly adding organic matter (compost or composted manure). This encourages biological activity in the soil.

Correct the soil pH as necessary. This can be accomplished by adding limestone.

Avoid overworking the soil by developing a crop rotation plan that leaves a portion of the land fallow (leave unplanted) every fourth year. The various soil components are removed by living organisms and are returned to the soil by death.

The bulk density of soil is influenced by soil structure due to its looseness or degree of compaction and its swelling and shrinking characteristics. Earthworms are probably the best-known soil organism that contributes to developing and maintaining soil structure.

Plant Nutrients

The three primary nutrients are nitrogen (N), phosphorus (P), and potassium (K). Together they make up the trio known as NPK. Other essential nutrients are calcium, magnesium, and sulfur.

· Molybdenum (Mo): Molybdenum helps bacteria
· Zinc (Zn): Zinc helps produce a plant
· Boron (B): Boron helps with the formation of cells
· Copper (Cu): Copper is an essential constituent

Plant nutrients in the soil

Soil is a significant source of nutrients needed by plants for growth. The three primary nutrients are nitrogen (N), phosphorus (P), and potassium (K). Together they make up the trio known as NPK. Other essential nutrients are calcium, magnesium, and sulfur. Plants also need small quantities of iron, manganese, zinc, copper, boron, and molybdenum, known as trace elements, because the plant needs only small amounts of these components for plant health. The role these nutrients play in plant growth is complex, and this document provides only a brief outline.

Major elements

Nitrogen (N)
Nitrogen is a critical element in plant growth. It is found in all plant cells, proteins, hormones, and chlorophyll.
Atmospheric nitrogen is a source of soil nitrogen. Some plants, such as legumes, fix atmospheric nitrogen in their roots; otherwise, fertilizer factories use nitrogen from the air to make ammonium sulfate, ammonium nitrate, and urea. Nitrogen is converted to mineral form, nitrate when applied to the soil so plants can take it up.
Soils high in organic matter, such as chocolate soils, are generally higher in nitrogen than podzolic soils. Nitrate is readily leached out of the soil by heavy rain, resulting in soil acidification. You need to apply nitrogen in small amounts often so that plants use all of it or in organic form, such as composted manure, to reduce leaching.

Phosphorus (P)
Phosphorus helps transfer energy from sunlight to plants, stimulates early root and plant growth, and hastens maturity.
The most common phosphorus source is superphosphate, made from rock phosphate and sulfuric acid. All manures contain phosphorus; manure from grain-fed animals is a particularly rich source.

Potassium (K)
Potassium increases plants' vigor and disease resistance, helps form and move starches, sugars, and oils in plants, and can improve fruit quality.
Also, heavy potassium removal can occur on soils used for intensive grazing and horticultural crops (such as bananas and custard apples).

Muriate of potash and sulfate of potash are the most common sources of potassium.

Calcium (Ca)

Calcium is essential for root health, the growth of new roots and root hairs, and the development of leaves. It is generally in short supply in the North Coast's acid soils. Lime, gypsum, dolomite, and superphosphate (a mixture of calcium phosphate and calcium sulfate) all supply calcium. Lime is the cheapest and most suitable option; dolomite is useful for magnesium and calcium deficiencies, but if used over a long period, it will unbalance the calcium/magnesium ratio. Superphosphate is useful where calcium and phosphorus are needed.

Magnesium (Mg)

Magnesium is a key component of chlorophyll, the green coloring material of plants, and is vital for photosynthesis (the conversion of the sun's energy to food for the plant). Deficiencies occur mainly on sandy acid soils in high rainfall areas, primarily if used for intensive horticulture or dairying. Heavy applications of potassium in fertilizers can also produce magnesium deficiency, so banana growers must watch magnesium levels because bananas are significant potassium users.

You can overcome magnesium deficiency with dolomite (a mixed magnesium-calcium carbonate), magnetite (magnesium oxide), or Epsom salts (magnesium sulfate).

Sulfur (S)

Sulfur is a constituent of amino acids in plant proteins and is involved in energy-producing processes in plants. It is responsible for many flavor and odor compounds in plants, such as the aroma of onions and cabbage.

Sulfur deficiency is not a problem in soils high in organic matter but leaches quickly. In coastal areas, sea spray is a significant source of atmospheric sulfur. The primary fertilizer sources are superphosphate, gypsum, elemental sulfur, and ammonia sulfate.

Main Functions of Plant Nutrients

Knowing the relative amounts of each crop's nutrient needs in making fertilizer recommendations is helpful. In addition, understanding plant functions and mobility within the plant should prove useful in diagnosing nutrient deficiencies. Supplying sufficient and efficient balanced nutrition is essential for a plant's optimal growth.

Yield

What Is Crop Yield?
Crop yield is a standard measurement of the amount of agricultural production harvested—yield of a crop—per unit of land area. Crop yield is the measure most often used for cereal, grain, or legumes; and typically is measured in bushels, tons, or pounds per acre in the U.S.
Sample sizes of a harvested crop are generally measured to determine the estimated crop yield for a larger region. Do not confuse yield with productivity, for productivity is the measurement of money produced per unit of land and not the total weight of a crop produced, which is yield.

What Factors Affect Yield?
Many of the factors already mentioned such as soil composition, soil fertility and macronutrients are all important factors that have an effect on crop yield. The importance of water and irrigation when it is needed will be covered in the following section. Other factors such as diseases and pests, weeds, climate and soil conservation have a great impact on crop yield. These factors need to be managed effectively to ensure increased productivity in large scale agricultural operations. As a mini-farmer, you will need to work to control these factors to produce the crops that you and your family will consume on a regular basis. This also includes the preservation and storage of crops for use after the harvest for year round consumption.

How Crop Yield Works
To estimate crop yield, producers usually count the amount of a given crop harvested in a sample area. Then the harvested crop is weighed, and the crop yield of the entire field is extrapolated from the sample.

For example, if a wheat producer counted 30 heads per foot squared, and each head contained 24 seeds, assuming a 1,000-kernel weight of 35 grams, the crop yield estimate using the standard formula would be 30 x 24 x 35 x 0.04356 = 1,097 kg/acre. Moreover, since wheat is 27.215 kg/bu, the yield we estimated would be 40 bu/acre (1097/27.215) or 40 bushels per acre.

Crop yield can also refer to the actual seed generation from the plant. For example, a grain of wheat yielding three new grains of wheat would have a crop yield of 1:3. Sometimes, crop yield is referred to as "agricultural output."

How Much Do You Need?

That is a really tough question to answer. It depends on your family size, how much each family member eats, what crops you are planting, whether you just want to eat fresh crops or produce enough to freeze or can for year round consumption. Analyze your grocery bills from the last few months to get a good idea of what you consume and use that as your guide as to what crops and how much of each crop to produce.

Watering and Irrigation

Irrigation is the artificial application of water to the soil through various systems of tubes, pumps, and sprays. Irrigation is usually used in areas where rainfall is irregular or dry times, or drought is expected. There are many irrigation systems in which water is supplied to the entire field uniformly.

Types, Methods, and Importance of Irrigation (Watering Crops)

Irrigation is watering crops, pastures, and plants using water supplied through pipes, sprinklers, canals, sprays, pumps, and other artificial features rather than purely relying on rainfall. In other words, it is an advanced watering system method for helping plants to grow because it is applied as an alternative to rain-fed farming.

It is also a technique of fulfilling plant or crop water requirements as they need it as an essential resource for growth. At the same time, it aids in providing plants with the nutrients required for development and growth and achieving high yields by enabling the penetration of roots in dry fields.

Types of Irrigation

1. Surface Irrigation
It is the most common type of irrigation as it simply employs gravity to distribute water over a field by following the contour of the land. In surface irrigation, for example, water will flow downhill from an area of higher elevation, reaching all the crops.
It is only applicable if the area or the land has sufficient water and is naturally sloped. Otherwise, it becomes very labor-intensive. It utilizes the furrow system technique, whereby channels direct water down a slope across a paddock where crops or plants are grown – about 1 meter apart.
The best example is rice paddies grown in East Asia. In those areas, the land is dug into terraces, and water flows downhill, allowing each plot of land to be watered.
However, surface irrigation is unsuitable for highly sandy soils with high infiltration, as it can lead to uncontrolled water distribution, resulting in floods and soil erosion. Also, it can only work in areas with an unlimited water supply.

2. Localized Irrigation
For localized irrigation, water is distributed under low pressure to each plant. Tubes or a piped network are used throughout the field, delivering water to each plant.

3. Drip Irrigation
Drip irrigation, also called trickle irrigation, is a sub-type of localized irrigation, where droplets of water are delivered directly to or near the roots of a plant at a very low flow rate.
It is an effective type of irrigation as it minimizes evaporation and water runoff. It is also very suitable for all types of topography and soils and perfect for areas with limited amounts of water or with high water costs. The pressure needed in drip irrigation is between 0.7 and 1.4 kg/cm2(10 and 20 psi).
Water is an essential element for survival. About seventy percent of the human body consists of water, while plants contain almost 90

percent of water. Still, we must depend on some outside sources to fulfill the water requirements of our bodies.

Similarly, crops require water for their growth and development.

Irrigation
Types
· Surface Irrigation
· Localized Irrigation
· Sprinkler Irrigation
· Drip Irrigation
· Centre Pivot Irrigation
· Sub Irrigation
· Manual Irrigation

Methods
· Traditional Methods
· Modern Methods
· Sprinkler System
· Drip System

What is Irrigation?

Irrigation is the process of applying water to crops to fulfill their water requirements artificially. Nutrients may also be provided to the crops through irrigation. The various sources of water for irrigation are wells, ponds, lakes, canals, tube wells, and even dams. Irrigation offers moisture required for growth and development, germination, and other related functions.

The frequency, rate, amount, and time of irrigation are different for different crops and also vary according to the types of soil and seasons. For example, summer crops require a higher amount of water as compared to winter crops.

Types of Irrigation

There are different types of irrigation practiced for improving crop yield. These irrigation systems are based on soils, climates, crops, and resources. The main types of irrigation followed by farmers include:

Surface Irrigation
In this system, no irrigation pump is involved. Here, water is distributed across the land by gravity.

Localized Irrigation
This system applies water to each plant through a network of pipes under low pressure.
Sprinkler Irrigation
Water is distributed from a central location by high-pressure overhead sprinklers or from sprinklers from the moving platform.

Drip Irrigation
In this type, drops of water are delivered near the roots of the plants. This type of irrigation is rarely used as it requires more maintenance.

Centre Pivot Irrigation
The water is distributed by a sprinkler system moving in a circular pattern.

Sub Irrigation
Water is distributed through pumping stations, gates, ditches, and canals by raising the water table.

Manual Irrigation

Manual irrigation is a labor-intensive and time-consuming system of irrigation. Here, the water is distributed through watering cans by manual labor.

Methods of Irrigation

Two different methods can carry out irrigation:
· Traditional Methods
· Modern Methods

Traditional Methods of Irrigation

In this method, irrigation is done manually. Here, a farmer pulls out water from wells or canals by himself or using cattle and carries it to farming fields. This method can vary in different regions.
The main advantage of this method is that it is cheap. But its efficiency is poor because of the uneven distribution of water. Also, the chances of water loss are very high.
Some examples of the traditional system are the pulley, lever, and chain pump. Among these, the pump system is the most common and used widely.

Modern Methods of Irrigation

The modern method compensates for the disadvantages of traditional methods and thus helps proper water usage.
The modern method involves two systems:
· Sprinkler system
· Drip system
· Sprinkler System
As its name suggests, a sprinkler system sprinkles water over the crop and helps in an even distribution of water. This method is much advisable in areas facing water scarcity.
Here a pump is connected to pipes that generate pressure, and water is sprinkled through nozzles of pipes.

Drip System

The water supply is administered drop by drop at the roots using a hose or pipe in the drip system. You can also use this method in regions where water availability is limited.

Importance of Irrigation

You can explain the importance of irrigation in the following points:
Insufficient and uncertain rainfall adversely affects agriculture. Droughts and famines are caused due to low rainfall. Irrigation helps to increase productivity even in low rainfall.

The productivity on irrigated land is higher as compared to the unirrigated land.

Multiple cropping is impossible in India because most regions have specific rainy seasons. However, the climate supports cultivation throughout the year. Irrigation facilities make it possible to grow more than one crop in most areas of the country.

Irrigation has helped to bring most of the fallow land under cultivation.

Irrigation has stabilized the output and yield levels.

Irrigation increases the availability of water supply, which in turn increases the farmers' income.

Irrigation should be optimum because even over-irrigation can spoil crop production. Excess water leads to waterlogging, hinders germination, increases salt concentration, and causes uprooting because roots can't withstand standing water. Thus the proper method is to be used for the best cultivation.

Irrigation Water Use

Center pivot irrigation systems are easily seen not only from space but from an airliner when you fly over the central United States. The circles might look tiny from the sky, but these devices can be huge down on the land surface, stretching thousands of feet.

Think of what your supper table might be like if you did not use water to irrigate crops. Could you survive very long without heaping servings of eggplant, beets, brussel sprouts, and rutabagas? Irrigation water is essential for keeping fruits, vegetables, and grains growing to feed the world's population, and this has been a constant for thousands of years.

Throughout the world, irrigation is the most critical water use. Estimates vary, but some say approximately 70 percent of all the world's freshwater withdrawals go towards irrigation uses. Large-scale farming could not provide food for the world's large populations without irrigation crop fields with water obtained from rivers, lakes, reservoirs, and wells. Without irrigation, crops could never be grown in California, Israel, or any deserts.

Irrigation has been around for as long as humans have been cultivating plants. After humankind learned how to grow plants from seeds, man's first invention was probably a hollow gourd to carry water. Ancient people labored intensively to carry water to pour on their first plants. Pouring water on fields is still a standard irrigation method today—but other, more efficient, and mechanized methods are also used. One of the more popular mechanized methods is the center-pivot irrigation system, which uses moving spray guns or dripping faucet heads on wheeled tubes that pivot around a central water source. The fields irrigated by these systems are easily seen from the air as green circles. There are many more irrigation techniques farmers use today since there is always a need to find more efficient ways to use water for irrigation.

When we use water in our homes or when an industry uses water, a large percentage of the water is eventually returned to the environment where it replenishes water sources (water goes back into a stream or down into the ground) and can be used for other purposes. It is estimated that of the water used for irrigation, only about one-half is returned to the environment. The rest is lost by evaporation into the air, evapotranspiration from plants, or is lost in transit by a leaking pipe.

Pest and Disease Control

Great battles are fought daily as gardeners strive to protect their precious crops while many garden pests and diseases seek to destroy them!
Pest and disease control presents a considerable challenge for the gardener, and identifying the cause of the damage is often the most challenging part! Those pesky pests don't generally show themselves, so how do you know whether a slug or a squirrel ate your plants? And just what is the difference between downy mildew and powdery mildew?
If you can't find a specific answer to your problem, here are eight general tips to help reduce the risk of attack.
1) Avoid using too much high nitrogen fertilizer as this promotes lots of soft leafy growth, particularly appetizing to garden pests.

2) Erect physical barriers. Covering crops with fleece or planting fruit in a fruit cage will often prevent pests from reaching your crops.
3) Know your enemy! Many pests and diseases are specific to particular crops. Find out what is likely to attack your crops and the type of damage that it will cause. Once you know what you are looking for, it is much easier to identify a problem when it occurs.
4) Keep your eyes peeled for the earliest signs of attack and take action. Don't let the problem get out of hand.
5) Practice good garden hygiene. Keeping your garden, greenhouse, and tools clean and well-maintained can prevent many diseases.
6) Improve your cultivation practices. Plants can become stressed by poor watering regimes, lack of nutrients, inadequate light and ventilation, or poor soil conditions. These stresses make them more susceptible from damage attacks by pests and diseases.
7) Encourage natural predators in the garden. Ladybirds and lacewings will happily munch their way through aphid colonies, preventing their spread throughout your garden.
8) If you need to use chemicals, then make sure that you read the label carefully and follow the manufacturer's instructions. Never spray chemicals near ponds and other water habitats. Always choose a still, calm day to undertake to spray. Always wear appropriate safety equipment.

Seed Starting

Seven Steps, from Seed to Garden
· Get the timing right. Have your seedlings ready to go outside when the weather is favorable.
· Find the proper containers.
· Prepare the potting soil.
· Start Planting.
· Water, feed, repeat.
· Light, light, light!
· Move seedlings outdoors gradually.

Get the timing right

The goal with seed starting is to have your seedlings ready to go outside when the weather is favorable. Start by looking at the seed packet, which should tell you when to start seeds inside. Usually, it says, "Plant inside six to eight weeks before the last frost."
Some vegetables, such as beans and squash, are best started outdoors. There is little benefit to growing them indoors because they germinate and proliferate. Some flowers, such as poppies, are best planted outdoors, too. These seeds are usually marked "direct sow."

Find the right containers

You can start seeds in almost any container, as long as they're at least 2-3" deep and have some drainage holes. If you like to do it yourself, you should grow seedlings in yogurt cups, milk cartons, or paper cups. I prefer the convenience of trays that are made especially for seed starting. It's easy to fill the trays, the watering system ensures consistent moisture, and I can move them quickly.

Prepare the potting soil

Choose potting soil that's made for growing seedlings. Do not use soil from your garden or reuse potting soil from your houseplants. Start with a fresh, sterile mix to ensure healthy, disease-free seedlings.
Before filling your containers, use a bucket or tub to moisten the planting mix. The goal is to get it moist but not sopping wet, crumbly, and not gooey. Fill the containers and pack the soil firmly to eliminate gaps.
Remember that most mixes contain few nutrients, so you'll need to feed the seedlings with liquid fertilizer a few weeks after they germinate and continue until you transplant them into the garden.

Start Planting

Check the seed packet to see how deep you should plant your seeds. Some of the small ones can be sprinkled right on the soil surface.

Larger seeds will need to be buried. For insurance, I plant two seeds per cell (or pot). If both seeds germinate, I snip one and let the other grow. Making a couple of divots in each pot is helpful to accommodate the seeds. After you've dropped a seed in each divot, you can go back and cover the seeds.

Moisten the newly planted seeds with a mister or a small watering can. Cover the pots with plastic wrap or a plastic dome that fits over the seed-starting tray to speed germination. This helps keep the seeds moist before they germinate. When you see the first signs of green, remove the cover.

Water, feed, repeat

As the seedlings grow, use a mister or a small watering can to keep the soil moist but not soggy. Let the soil dry slightly between watering. Set up a fan to ensure good air movement and prevent disease. I use a fan plugged into the same timer as my grow lights. Remember to feed the seedlings regularly with liquid fertilizer, mixed at the rate recommended on the package.

Light, light, light!

Seedlings need a considerable amount of light. If you're growing in a window, choose a south-facing exposure. Rotate the pots regularly to keep plants from leaning into the light. If seedlings don't get enough light, they will be leggy and weak. If you're growing under lights, adjust them, so they're just a few inches above the tops of the seedlings. Set the lights on a timer for 15 hours a day. Keep in mind that seedlings need darkness, too, so they can rest. As the seedlings grow taller, raise the lights.

Move seedlings outdoors gradually

It's not a good idea to move your seedlings directly from the protected environment of your home into the garden. You've been coddling these seedlings for weeks, so they need a gradual transition to the great outdoors. The process is called hardening off. About a week before you plan to set the seedlings into the garden, place them in a protected spot outdoors (partly shaded, out of the wind) for a few hours, bringing them in at night. Gradually, over the course of a week or ten days, expose them to more and more sunshine and wind. A cold frame is a great place to harden off plants.

A special tray for starting seeds seed starting tray with a built-in watering system makes seed starting foolproof. Potting soil is a substrate that's made especially for seed starting. Good-quality "potting soil" for seed starting doesn't have any soil in it. This sterile, free-draining mix is perfect for seeding.

Starting Seeds Indoors: A Step-by-Step Guide
Who Should Start Seeds Indoors?

Starting seeds indoors under lights is quite the process, but don't worry. It's simple when you follow these instructions, and it gets easier with practice!

However, there is an initial investment to acquire the equipment and materials, so if you're new to gardening or super busy, it's better to skip this step.

Instead, you can start seeds out in the garden when the weather warms up (for some crops) and buy seedlings from a local farmers' market or nursery for those crops that need more time to develop.

Some gardeners start seeds in south-facing, sunny windows rather than under lights. Still, success with this method can be inconsistent for various reasons, such as lack of daylight hours, extreme temperature fluctuations, and a reduction in UV light by the windows themselves. Modern windows may only transmit 25% of ultraviolet light, while older, plain-glass windows may only transmit 80% of light.

Check out this comparison of seedlings grown on a windowsill vs. under lights.

Heat mats can be beneficial if your budget doesn't allow starting seeds under lights just yet and you intend to use a windowsill.

Ultimately, you won't find two gardeners who start seeds indoors precisely the same way, but the following is the system that has worked for me for many years.

When to Start Seeds Indoors

For most people, your frost dates determine when to start seeds.
Once you know your frost dates, you can set a schedule for starting
seeds indoors using my article When to Start Seeds: Your Guide to
Spring Planting. If you're starting seeds for fall, check out When to
Start Seeds: Your Guide to Fall Planting.
You may also want to check my Year-Round Gardening Guide,
which details to-do lists for each month.

Get Started #1: Set up Shelves

The shelving unit you choose determines many things, such as the
size of your lights and how much of the other equipment listed
below you will need.
You can use wall-mounted wire shelves. Throughout the year, these
shelves act can be used as a root cellar where harvested vegetables
such as garlic, onions, sweet potatoes, and winter squash can be
stored. After the New Year, you can condense your stored vegetables
and make room for starting seeds on these shelves.
If you need more than wall-mounted shelves, try building your
wooden shelves or purchase a free-standing, 6-tier wire shelving
unit.
The wire shelving model of preference has the shelf space to hold
two standard nursery trays (22″). Find a unit that can hold ten
nursery trays, double what other 6-tier shelving units can hold. Just
be sure to compare the height of your ceiling to the dimensions of
the shelving unit before purchasing. It's easiest if you can access
both sides of the shelves, which is why the free-standing units are
recommended. However, you can manage fine with single-side
access to the wall-mounted shelves. Starting seeds indoors under
lights can seem intimidating. However, this step-by-step guide
demystifies the seed-starting process and helps you get started
confidently.

Get Started #2: Select Lights

The type of lights to use for indoor seed starting is a topic of great debate. Many experienced growers swear by inexpensive shop lights, while others insist that the best quality comes from specialized grow lights.

In reality, there are three reasons why this debate continues.

Firstly, there's a difference between the needs of commercial growers growing exclusively under lights and home gardeners who grow seedlings to a specific size before transplanting them outside.

Secondly, lighting technology is constantly changing, so a model that may have been expensive and inconsistent a few years ago may now lead the pack in affordability and reliability.

Thirdly, the science concerning lighting could be more technical and often needed to be understood.

If you already have fluorescent shop light ballasts, then full-spectrum fluorescent tube lights should be sufficient. Be sure the ballasts fit the width of your shelving unit. Typically you will need two ballasts, side by side, for each shelf; 32-watt T8 bulbs work well.

Go ahead and keep using those old shop lights if that's what you have. However, replace the bulbs at least every other year to keep sufficient brightness for growing.

You'll also need screw hooks for wooden shelves OR s-hooks for wire shelves (2 per fluorescent light fixture) to hang the lights from the shelving unit.

Are LED Grow Lights Better?

LED grow lights are still more expensive than the old fluorescent shop lights, but they've come a long way. To date, they generally provide a superior seed starting experience and are a better investment if all the equipment you're buying is new.

LED grow lights now mimic sunlight more closely, with smaller and lighter units that are extremely energy efficient and long-lasting. They even have white-yellow lights.

You would need two of these lights for each shelf of the 6-tier unit recommended above, which equals ten lights in total. This is a significant expense, so start with your equipment and add/improve over time.

Get Started #3: Select Seed Starting Mix

There are two significant issues to consider when selecting a seed starting mix: sustainability and herbicide contamination.

Let's take a brief look at each.

Concerning sustainability, there are two materials widely used as the main ingredients in bagged soils and soilless mixes: peat moss and coconut coir.

There has been an environmental issue concerning peat bogs being destroyed in Canada to cultivate peat moss for gardening products and coconut coir has emerged as a "sustainable" alternative.

Although it isn't a 1:1 replacement (peat is superior in performance), coir works pretty well.

Unfortunately, the coir industry also has its' environmental impact and pollutes and depletes the environment where it's made.

In addition to the peat vs. coir debate, many bagged soils and potting mixes contain compost as the main ingredient and it can be contaminated with herbicide residue.

This residue can damage your seedlings and impact your garden for years.

If peat moss, coconut coir, and compost are all "bad," what can you do?

First, remember that you have the most control over what you do on your site.

Learn about how herbicides can contaminate your garden and compost (even if you don't use herbicides yourself), and seek to keep them out. Then use compost created in your garden for seed starting mix as well as improving the soil in your garden.

In addition, rather than throwing old potting soil in the compost or trash, consider reusing it as a sterilized seed starting mix. Here are some tips for sterilizing potting soil and compost for seed starting.

The best way to sterilize your seed starting mix for small amounts is steaming or cooking. Use a pressure cooker at 180 degrees for 30 minutes, bake in the oven at the same temperature/time.

Alternatively, you can use the microwave. Use the solar method for raised beds. Wet the soil, cover with clear plastic weighted down on the edges to cover and leave for 4 to 6 weeks to thoroughly sterilize the soil.

Second, buy a seed starting medium that you feel is the best choice, using (and reusing) it wisely.

Don't be fooled by bagged soils that have "Organic" on their label without indicating that they've been verified and approved for organic agriculture by a third party.

If a seed starting mix contains compost, rigorously interview the company about their practices to avoid herbicides. Tilth Soil's "Sprout" seed starting mix is approved for organic agriculture and can be used for both seed starting and potting soil for transplants. Currently, coconut coir is the most widely available main ingredient for a seed starting mix that is approved for organic agriculture.

Your Step-by-Step Guide for Starting Seeds Indoors

Step 1: Assemble the shelves.

Assemble the shelving unit and set it in a level spot in a room with as much ambient light as possible. My seed starting room is in the basement with no ambient light, so I cover the walls with aluminum foil (shiny side out) to reflect the light. Mylar reflective film is less flimsy, while a large mirror can work, too.

Step 2: Hang the lights.

Hang the lights using screw hooks or s-hooks, and mount a clip fan at the end of each shelf.

Step 3: Manage the temperature and humidity.

Set the heat mats and the thermometer on or near the shelving unit. The ideal soil temperature for starting seeds indoors is 68-86° F, usually attainable with ambient heat in the house and the supplemental heat mat. The accompanying thermostat can ensure you keep your seedlings within a healthy temperature range for germination.

If your seed starting operation is in a small, enclosed room, you can use an extra thermometer and humidity reader to ensure good conditions. A small space heater warms the air in the enclosed space. As for humidity, seedlings prefer 50% to 70% humidity. Cool indoor air is usually dry, so keep a spray bottle of water in the room and mist the air several times per day.

Step 4: Manage the power source.

Mount the power strip and organize the cords with twist ties so they're out of the way for regular seedling maintenance and away from possible water spillage.
Plug the light fixtures into the timed outlets and the fans and heat mats into the always-on outlets.

Step 5: Ready to start seeds? Assemble the materials.

Collect your trays, cell packs, seeds, seed starting mix, steel tub, plant markers, permanent marker, watering can (filled), garden gloves, and pencil near your work table.
Feel free to use materials that you already have around the house. For example, you can root through your kitchen and use metal roasting pans as drainage trays.
Plastic food-safe containers that you usually recycle are just fine to use as either drainage trays or seed starter pots, depending on their size.

Step 6: Prepare the planting medium.

Fill the tub with seed starting mix and add water, one gallon at a time, mixing just until the soil medium clumps together. It should feel like a wrung-out sponge: If you can squeeze water from it, it's too wet. To remedy this, add more soil.
You can save excess planting medium. Let the medium dry out before storing it when you're finished starting the seeds.

Step 7: Prepare the cell packs.

Fill each cell of a cell pack to the top loosely with the seed starting medium. Pack it firmly so that each cell is only one-third to 2/3 full. Then fill each cell again loosely to the top. This time, press lightly rather than pack down firmly.

Step 8: Choose a seed packet to begin starting seeds indoors.

Select a seed packet to work with and consult your garden plans to figure out how many seeds to start. Always start a few extra, just in case.

Step 9: Plant the seeds.

Plant seeds twice as deep as their size, at least two per cell in opposite corners. Use the pencil to push the seeds in, then lightly press the soil on top, so the seeds are covered.
Press tiny seeds into the top, like lettuce seeds, rather than covering them.

Step 10: Label the cell packs.

Label each cell pack with a plant marker.

Step 11: Fill the drainage tray with cell packs.

Place the cell packs in a drainage tray (the nursery tray without holes).

Step 12: Water the seeds.

When the drainage tray is full of cell packs, water each cell very lightly from the top to ensure the seeds have made contact with the soil.

Step 13: Cover the trays for germination.

Cover the tray with plastic wrap or a clear germination dome and set it on a heat mat. This keeps seeds warm and moist to increase germination rates.

seedling pots covered with plastic wrap to improve germination Cover newly seeded trays with plastic wrap or germination domes and set them on heat mats. Turn lights off until seeds have germinated.

Step 14: Start the rest of your seeds.

Continue the process until all seeds have been started and all trays have been covered and set on a heat mat.

Step 15: Turn on the heat mats.

Make sure all heat mats are on. Lights and fans are OFF until the seeds have germinated.

Step 16: Check daily for germination.

Each day check for germination, briefly lifting the plastic wrap or dome to allow some condensation to escape. Once seeds have sprouted, immediately remove the plastic covering.

Supply List for Starting Seeds Indoors

In addition to shelving, lights, and seed starting mix, the following supplies make the whole process run smoothly. You won't have to buy most of these items ever again! Check out my one-stop Amazon shop to see all this equipment in one place.
Aluminum foil, Mylar reflective film, or mirror for low light areas
Clip fans (one for each shelf)
Seedling heat mat with thermostat (1-2 per shelf)
Thermometer/humidity monitor
Programmable power strip
Waterproof table OR shop-style workbench
Standard nursery growing trays (no holes – 22" x11")
Plastic seed starter cell packs OR reduce plastic with a soil blocker + nursery trays (with holes)
· Seeds (I check Botanical Interests first.)
· Large galvanized steel tub
· Plant markers

· Permanent marker
· Small watering can
· Garden gloves (optional)
· Pencil (sharpened)
· Plastic wrap (Saran-style) OR plastic germination domes to fit nursery trays
· Potting soil: Tilth Soil's "Sprout" (all-in-one seed starting and potting soil mix) or Nature's Care potting soil
· Plastic pots
· Cinnamon spice shaker (optional)
· Fish & Seaweed Fertilizer
· Garden Scissors

Selecting and Saving Seeds

Traits you should look for when you consider saving seeds for future plantings.

· Vigor. This is the seed's ability to germinate quickly and grow into a healthy, productive plant.

· Earliness.

· Trueness-to-type.

· Disease and Insect Resistance.

· Tolerance to Drought or Excess Moisture.

· Stockiness.

· Hardiness.

· Lateness to Bolt

SEED SAVING: SELECTING PLANT CHARACTERISTICS

Seed ordering, as we know it today, is a recent development in the history of agriculture. For millennia, seed preservation was a routine part of gardening. This family practice led to the creation of heirloom varieties and the adapting of crops to suit local conditions.

Whenever we save seeds, we decide what future generations of plants will look like. For that reason, saving seeds for next year takes careful consideration. That is especially true with easy-to-save species like potatoes or garlic, and it can be tempting to save and plant whatever is leftover, or you don't want to cook with. Still, it's important to remember that this will affect future harvests. Take a look at some traits you should consider when saving seed.

Vigor

This is a seed's ability to germinate and vigorously grow into a healthy and productive plant.

Earliness

Regardless of the length of your growing season, you should select for early production. Plants that mature quickly allow for earlier produce, pest avoidance, or multiple harvests.

Trueness-to-type

If you're trying to preserve an heirloom variety, save only seeds from plants that are true to type and display only the unique characteristics.

Disease and Insect Resistance

Saving seeds every year can adapt a variety to withstand your local pests and diseases better.

Tolerance to Drought or Excess Moisture

This trait is another way to adapt a variety to your garden location.

Stockiness

Tall, spindly plants tend to lodge against neighboring plants and other problems, often requiring additional trellising. On the other hand, stocky plants are often healthy plants. Nutrient availability and how closely you space your plants can also affect this trait.

Hardiness (Cold Resistance)

Crops grown in early spring or late fall need to be hardy, and you want to select seeds from plants that withstand cold temperatures.

Lateness to Bolt

Avoid saving seeds from the first plants that go to seed. Selecting seeds from plants that bolt later in the season will increase your harvest period.

Color

This trait does not impact the crop yield and, therefore, may not seem as important as disease resistance or flavor, but often, it is the unique color that makes people fall in love with an heirloom variety. Selecting seeds with the most intense purple color of Purple Dragon Carrots is part of what makes this variety unique.

Uniformity

This trait depends upon what you want from the variety in question. You may want all of your pea plants to be a uniform height, but for obvious reasons, you don't want your rainbow Swiss Chard to have a uniform color.

Flavor

Heirloom vegetables are best known for their delicious flavor. Remember this when saving seeds for future crops, as no one wants to eat bland vegetables.

Flesh Characteristics

This trait depends mainly on a variety's purpose. For example, tomatoes bred for drying, such as Principe Borghese, should have much less moisture than those bred for slicing, such as Radiator Charlie's Mortgage Lifter.

Size and Shape

If you love stuffed jalapenos, save seeds from the giant peppers. This is another excellent way to make a variety of work for you.

Productivity

Please don't eat your biggest, best cabbages even though it's tempting. Let those go to seed so that more of your cabbages resemble the best in a few years.

Storage Ability

With modern fridges and freezers, storage may seem less important. However, it's still vital to consider storage ability, especially for plants like storage tomatoes, winter squash, sweet potatoes, and pumpkins.

When selecting the right seeds, it can be overwhelming to consider all the different aspects. To make things easier and keep track of desirable traits, you can try tying a brightly colored piece of yarn loosely around a specific plant. This will help you remember which plants displayed traits like vigor when it's time to harvest seed. However, it would be best to keep an eye on the yarn as the plant grows so that it doesn't become too tight and harm the plant. Alternatively, you can place small stakes in front of plants to mark them.

Season Extension

What exactly is a season extension? Season extension refers to the agricultural practice that allows a crop to be cultivated beyond its typical outdoor growing season.

Advantages:

Possible year-round income

Retention of old customers

A gain in new customers

Higher prices

Higher yields

Better quality

Extended employment for workers

Disadvantages:

No break in the yearly work schedule

Increased management demands

Higher production costs

Plastic disposal problems

How does season extension contribute to sustainability? "…to make a real difference in creating a local food system, local growers need to be able to continue supplying "fresh" food through the winter months…[and] to do that without markedly increasing our expenses or our consumption of non-renewable resources". – Eliot Coleman, The New Organic Grower Thermodynamics and Properties of Plants.

Extending the growing and harvest seasons of crops is essential, and we can achieve it by using techniques that focus on two primary goals: protecting crops from extreme weather conditions and enhancing their growth for quicker maturity and better quality. Often, one approach can serve both purposes. For instance, a raised bed can dry faster and warm up quickly in spring, but it requires more attention to irrigation needs and may develop higher soil temperatures sooner than flat ground in summer. On the other hand, a row cover can protect crops from frost, but it can also prevent the crop from developing as hardy as an uncovered crop due to the mild artificial climate under the cover. Therefore, it's crucial to understand some basic principles about heat and cold, how plants respond to thermal changes, and how various landscape features and protective materials influence the thermal environment of plants to extend the crops' growing season successfully.

Controlling the flow of heat: The ground is a massive reservoir of heat. The heat radiating from the soil protects crops at night when you cover them or protects the lower part of plants with a good canopy of leaves. As the ground gets colder in late fall, it radiates less heat. Thus, the leaf canopy or added row covers will give less protection as the ground cools. The reverse is true as the soil temperature increases in spring. Wet ground conducts and radiates more heat than dry ground. Lighter, sandier soils and soil in a raised bed will dry out, warm up, and become workable earlier in the spring. Adding organic matter to clay soils can improve drainage; this will also darken any soil so it will absorb more heat. Dark mulches, as well as black plastic, also raise soil temps. Mulches will insulate the ground and allow less heat out at night; thus, crops will be colder on the mulched ground.

Clouds form a "blanket" that slows radiant heat loss from the earth. Temperatures can drop as much as 5-10° F within an hour after the sky clears at night. The arrival of cloud cover will often raise temperatures and save crops on a frosty night. Water can store a large quantity of heat and release it relatively quickly. This enables overhead irrigation to protect a crop from frost. It also creates warmer microclimates near ponds or other bodies of water. Water absorbs heat quickly, and evaporating water removes heat from its environment. Thus, overhead irrigation on a hot day can cool heat-sensitive crops or aid the germination of heat-sensitive seeds. Land sloping to the south will stay warmer in the late fall and warm up sooner in the early spring. Land sloping east will warm sooner in the morning; to the west, it stays warmer in the evening. Cold air is heavier than warm air and will slide down slopes to settle in flat areas or hollows, often called "frost pockets." On a calm night, the warmest "microclimate" in a given area will often be near the top of a slope.

Objects in a landscape, such as buildings or artificial windbreaks, can influence the movement of cold air. Windbreaks uphill from a crop can protect from frost; downhill from a crop may cause trapping of cold air. A forest surrounding a smaller, level field on several sides may also keep cold air from reaching the crops. Windbreaks, such as buildings, hedges, fencerows, or woods, can influence microclimates in other ways. They can create an advantageous "microclimate," where solar gain can accumulate in the daytime; wind can stress plants by accelerating leaf evaporation and pushing cold air more profoundly into the crop's canopy or through a row cover. The "wind chill factor" means: moving air extracts heat from objects faster than still air.

Season extension is a technique that farmers use to cultivate or harvest crops outside their conventional growing season. By using these techniques, growers can gain a more significant market share by providing products to retailers when their competitors cannot. The primary goals of season extension are to protect crops from frost or heat damage and to speed up crop growth for quicker maturity and higher quality.

Farmers have developed various season extension techniques to produce crops earlier in the spring, grow cool-season crops in summer, maintain production in the fall, and even harvest crops in winter. These practices range from simple ones like planting early-maturing, cold-hardy, heat-tolerant varieties, planting windbreaks, or irrigating crops to reduce damage from extreme temperatures to more complex ones such as year-round production in heated greenhouses.

Over centuries, farmers have used several time-honored methods, such as using cold frames heated with manure, shade structures, windbreaks, irrigation, masonry walls, or stone mulch as heat sinks and cloches to protect individual plants.

The use of plastic in horticultural crop production has seen a significant increase in recent years, extending the possibilities for year-round production. Farmers now use variously colored plastic film mulches, row covers, shade cloths, low tunnels, and high tunnels/hoop houses to protect crops from weather-related damages. High tunnels have become increasingly important to market gardeners and are being established nationwide.

Examples of Season Extension Techniques:

The selection of crop varieties that mature over a range of dates benefits farmers, allowing them to harvest crops over an extended period. Heat-tolerant varieties can be grown during warm months, and cold-tolerant varieties can be grown during cold months, stretching out the proportion of the year in which the crops can be sold.

In many regions of the U.S., salad greens and cole crops can be successfully grown most of the year if appropriate varieties are selected. This can be combined with other season extension strategies such as shade cloth and high tunnels. Conducting on-farm variety trials is an excellent way to identify the best-performing varieties in a particular region and when grown under a farm's unique suite of practices.

Raised beds are planting beds where the soil is loosened and piled up to a level above the surrounding soil surface. By doing so, they heat up more quickly in spring, allowing for earlier planting. Mulches are materials placed on the soil around plants for various purposes. Plastic mulches, usually plastic sheeting with slits through which plants grow, are extensively used in large-scale vegetable production to suppress weed growth, retain soil moisture, increase soil temperatures, and speed up crop growth. Organic mulches, on the other hand, retain soil moisture, increase organic matter content, and cool soils.

Plastic mulches are permitted for use in organic production systems as long as the plastic is removed from the field at the end of the season. Row covers, which are light, porous, permeable fabrics placed over plants, retain heat and protect from wind and insect pests. They also offer several degrees of frost protection. Cold frames, transparent-roofed enclosures built low to the ground, protect plants from cold weather and are primarily used for growing seedlings that will later be transplanted into the field. High tunnels, which are metal frames covered in plastic sheeting, function similarly to greenhouses but are generally unheated and, in most cases, do not have exhaust fans. Many high tunnels are constructed to be moved from one location to another to permit crop rotation and soil management.

Fully heated and artificially lit greenhouses are the ultimate season extension device, allowing crops to be grown year-round, even through sub-zero winters. The adoption of this energy-intensive form of season extension by organic farmers has been debated in organic and sustainable agriculture communities.

Irrigation is applied to crops to prevent damage due to high or low temperatures. Wind machines move air to prevent the accumulation of cold air near the ground, protecting blossoms and crops from cold damage.

Fruit Trees and Vines

10 Fruit Trees and Edible Vines for Your Garden You Might Not Know

Fruits are more than just the standard varieties we usually find in supermarkets. The world of fruit is far more expansive and exciting than we think. Have you ever tasted a translucent white mulberry or jujube, a fruit and not a candy? These are just two examples of fruits that are not commonly sold in groceries but can be found in local grocers or farmers' markets. It's not because they're difficult to grow; most of them are way easier to raise at home than other fruits like peaches and cherries. They're not commercially available because they score low on metrics such as yield per acre and shippability. However, this will be fine if you grow them in your backyard. A tree can yield more than enough for your family, and the fruit must only travel a short distance from the garden to the kitchen. If your local garden center needs to have the varieties you want, you can easily find them online from mail-order nurseries.

Here are ten unique fruits that you might not have heard of before:

1. White Mulberry: The fresh fruit of this small, attractive tree has a less acidic flavor than its dark-colored counterpart. Its dried version is sold in health stores as a "superfood," but can be expensive. Several white-fruited mulberry varieties are available, including Tehama, Beautiful Day, and Sweet Lavender. The tree grows in USDA zones 4-9 and is often used to cultivate silkworms in Asia.

2. Jujube: This Chinese fruit is consumed dried, which gives it a chewy, candy-like texture. It has a sweet-sour flavor and grows on thorny trees with a narrow, upright growth habit. Jujube trees are highly drought-tolerant and thrive in hot, dry areas. They grow in USDA zones 5-9.

3. Cider Apple: Heirloom apple varieties can be used to make cider, but real cider makers use unique varieties bred for centuries with a flavor profile suited to the beverage. These varieties include Ashmead's Kernel, Northern Spy, and Muscadet de Dieppe. If you like home-brewed cider, you should grow your own. These apples grow in USDA zones 4-9.

4. Pawpaw: This little-known native fruit is found in isolated patches throughout eastern forests. It grows on small, slow-growing trees with attractive foliage and a uniform pyramidal shape. The fruit is the size of a mango and has an exotic flavor often described as a cross between banana, pineapple, and mango. Pawpaws are far too demanding for commercial growers, but they've garnered a cult-like following among foodies and backyard botanists. They grow in USDA zones 5-9.

5. Pineapple Guava: The fruit of this small, attractive evergreen tree tastes like pineapple-flavored guava. Its large red-and-white tropical blossoms are also edible, adding a sweet, cinnamon-like flavor to desserts and summer drinks. Pineapple guavas, also known as feijoas, are not cold-hardy. You can grow them outdoors year-round in much of California, southern Texas, Florida, and the Deep South. Potted pineapple guavas are easily maintained as small shrubs. They grow in USDA zones 8-11.

6. Quince: This fruit is related to apples and pears, and in past centuries, it was just as popular in northern European households. Quinces have a bloated, tumor-laden pear-like appearance and require cooking to be enjoyed. However, their flavor is nonpareil, like a baked apple with cinnamon, allspice flavors, and a touch of lemon zest. They grow in USDA zones 4-9.

7. Loquat: This fruit is a distant relative of apples and pears from subtropical parts of Asia. It looks like an apricot, with a similar

texture and flavor but tangier. Loquat trees are evergreen and require a warm climate. While they're not giant trees, they are a bit large to grow in pots and bring indoors for winter. They grow in USDA zones 8-10.

8. Arctic Kiwifruit: This variety of kiwifruit hails from the frigid mountains of Russia and has a flavor similar to fuzzy kiwifruit but lacks fuzz. Arctic kiwifruit is typically consumed whole, skin and all. This shade-tolerant vine has spectacular white, pink, and green-variegated foliage. It grows in USDA zones 3-8.

9. Chocolate Vine: Also called akebia, this shade-tolerant vine has delicate lobed foliage and bears vanilla-scented flowers in spring. In summer, sausage-shaped pods appear, which split open when ripe to reveal a soft, white pulp flavored with notes of banana, lychee, and passion fruit. The pod is inedible raw but may be cooked like a vegetable. You can find this vine in USDA zones 4-9.

10. Maypop: The passion fruit is the American cousin of its subtropical counterpart, which requires a subtropical climate. Maypop vines are nearly identical to their tropical counterparts, with frilly purple and white blossoms up to three inches in diameter. You can mix the yellow flesh of the fruit in smoothies, daiquiris, and desserts. The leaves of maypop are considered an herbal aphrodisiac. They grow in USDA zones 6-10.

Vines You Should Grow in Your Fruit Tree Orchard

As you plan for the upcoming gardening season, consider adding vertical garden vegetables to your layout. Growing climbing vegetables can help you make the most of your space by utilizing posts, trellises, and fences to grow upwards. Though most vegetable vines are annuals, a few perennial vining veggies can be grown year-round in temperate or tropical zones.

Raising Chickens For Eggs

Raising chickens can be a thrilling experience, but it requires some effort. Chickens need to be fed regularly and provided with clean water every day. They should also be removed from their coop every morning and put back in at dusk to protect them from predators. Collecting eggs twice a day and cleaning the coop and pen once a week is recommended to maintain hygiene and prevent unpleasant odors. This guide is aimed at beginners who want to learn how to raise chickens properly.

Why Should You Raise Chickens?

Raising chickens in your backyard can be an excellent idea for many reasons. The eggs are much fresher and tastier than store-bought ones and are perfect for baking. You can also compost the shells and chicken poop, which is great for your garden. Chickens are also extraordinary at picking at grass, worms, beetles, and other insects, making them perfect gardening companions. They are pretty independent and can keep themselves entertained most of the day.

Things to Consider Before Getting Chickens

Before investing time and money into preparing for chickens, it's important to check local town ordinances to ensure that keeping chickens is allowed in your neighborhood and that there is no limit to the number of chickens you can have. It would be a disappointment to find out that you can't keep chickens after all your preparations!

Ensure you have enough space for a henhouse or a full-size chicken coop. The coop should be large enough to hold a feeder, water containers, a roosting area, and a nest box for every three hens. It should be sturdy enough to keep your chickens safe from predators. A proper coop should allow you to comfortably gather eggs and shovel manure, though a simple henhouse can be smaller.

If you are planning to raise chickens, it is crucial to remember that they need to be fed and given water daily. The cost of feed is around $20 per 50-pound bag at a local co-op, but this price may vary depending on your location and the quality of the feed. The duration of the feed bag depends on the number of chickens you have.

Hens are known to lay eggs during spring, summer, and fall as long as they receive 12 to 14 hours of daylight. You can expect to collect eggs daily or even twice a day.

Cleaning up manure is a year-round task. If you plan to go on vacation, you will need a reliable chicken sitter, which can be challenging.

Flock size, spacing, and start-up cost are essential factors to consider when raising chickens. Chickens are sociable creatures and require company, so keeping at least three to six birds together is recommended. By doing so, you can expect a steady supply of eggs since an adult hen lays around two eggs every three days.

Notably, chickens are most productive in the first two years of their lives. After that, their egg production declines, and you may need to replace them with younger birds. You can easily buy young chicks from suppliers or hatch your own if you have a rooster.

The amount of space chickens need depends on the breed you raise. According to the University of Missouri Extension, one medium-sized chicken needs at least 3 square feet of floor space inside the coop and 8-10 square feet outdoors. The more space you provide, the happier and healthier your chickens will be; overcrowding contributes to disease and feather picking. Chickens need a place to spread their wings, such as a sizeable chicken run or a whole backyard. The space must be fenced to keep the chickens and predators out. (Predators include your own pets too!) Make sure to add chicken-wire fencing to your list of equipment.

How Much Does Keeping Chickens Cost?

Raising chickens can be a costly endeavor. Building a coop and run that is 20x5-feet in size, complete with wood, fencing, and hardware, can cost you around $300 or more. If you need to gain the skills to do it yourself, you'll also need to pay for skilled labor.

All in all, expect to spend anywhere from $500 to $700 when starting out, depending on the size of your flock, coop, and run.

Gardening with Chickens

Many people keep chickens for the sole purpose of having a steady supply of fresh eggs. However, did you know that they can also benefit your garden? When the gardening season is over, you can let your chickens into the gardening area. They will eat any damaged or overripe vegetables, weed seeds, or insects found in the soil. They will also scratch the ground, peck hidden worms or insects, and mix up the soil.

Additionally, chickens produce endless manure, which can be composted, aged, and added to the garden. You can collect and pile the chicken poop and used bedding materials during your daily coop cleaning and add other materials such as lawn clippings, fruit and vegetable kitchen scraps, leaves, twigs, and shredded paper. Soak the pile, wet it, and stir it regularly to add air. A temperature of 130°F to 150°F is recommended to eliminate bacteria.

If you're wondering how to collect eggs from a chicken coop, our article on collecting, cleaning, storing, and hatching chicken eggs has all the answers. Once you've tasted farm-fresh eggs, you'll never want to go back to store-bought eggs. Farm-fresh eggs are delicious, with bright yolks and firm whites, while grocery-store eggs are often up to a month old before they even get to the stores. If you're planning to raise chickens, having a steady supply of fresh eggs is your primary reason for getting the birds in the first place!

How Often Do Chickens Lay Eggs?

Hens lay about one egg per day when they're laying. You'll collect more eggs during extremely warm or cold weather as the hens spend more time in their coop. Collecting eggs frequently keeps the eggs from breaking due to hen traffic. Always discard eggs with cracks, which allow bacteria to enter the egg.

Also, be sure the shells are strong. Give your hens ground oyster shells or a similar calcium supplement, available at farm suppliers, to promote the development of strong eggshells.

How to Collect Eggs

When collecting eggs from chickens, waiting until they leave their laying spots is best. You can feed them before collecting the eggs to encourage them to move. Make sure you use a basket, a cloth sling, or any other container that won't exert pressure on the eggs. However, there may be times when a hen becomes broody and refuses to leave her nest. If this happens, you may have to remove the eggs from under her. Be prepared for some squawking and maybe even a few indignant pecks!

When to Collect Eggs

You should collect eggs every morning, as hens cackling loudly is a sign they're laying. Check again in the evening as some hens lay in the morning and others at night.

Why Are My Chickens Eating Their Eggs?

It's interesting to note that chickens have a taste for eggs just like we do. In fact, many chickens develop an appetite for eggs after trying them out from broken shells. Chickens are known to be opportunistic eaters and will peck at anything that looks like food. To avoid egg-eating, it's important to immediately clean up any broken eggs and discard any "eggy" straw or shavings. Once a chicken learns this habit, it's challenging to break it, and other chickens may follow suit. It's essential to avoid having your chickens consume your eggs; you want to be able to enjoy them yourself!

What Color Eggs Will My Chickens Lay?

Did you know that the color of a hen's ear can tell you what color eggs she will lay? Instead of external ears, birds have a small circle or oval skin on the side of their head next to the ear hole. If this area is white, the hen will lay white eggs. If it's red, she will lay brown eggs. Although there is no difference in flavor or nutrition, white eggs are better for dyeing at Easter.

When you have farm-fresh eggs, avoid washing them if possible. Instead, wipe them with a dry, rough cloth. The eggshell has a natural coating called a "bloom" that protects it from bacteria. If you wash the eggs, this protective layer will be removed, and the eggs must be refrigerated. However, if left unwashed, they can be stored on the counter for up to a month or in the refrigerator, depending on your personal preference. Eggs are best consumed within two weeks but can be eaten up to a month after they have been laid.

If the eggs have manure on them, be sure to remove it. To keep your eggs clean, keep the straw fresh and remove any large pieces of debris. However, some eggs may inevitably have a bit of dirt on them. In this case, wipe them gently with a damp cloth. You can submerge and scrub them with a vegetable brush for filthy eggs. Always use warm water warmer than the egg; cold water can cause the egg to shrink inside the shell and draw in bacteria.

If you do wash the eggs, be quick and gentle. After washing, let the eggs air-dry entirely before storing them. You can store your eggs in dated egg cartons on a shelf in the refrigerator, but avoid placing them in the door where they can be jostled with every opening and closing. For partially filled cartons, mark each egg with the date it was collected using a pencil. Keep the eggs refrigerated between 32 and 40 degrees Fahrenheit. Remember, fresh eggs can be stored in the refrigerator for up to a month.

Hatching Chicken Eggs

Many people wonder whether a chicken can hatch from an egg purchased at the grocery store. Unfortunately, the answer is no. A chicken must first be fertilized for it to develop from an egg. Most eggs sold commercially in grocery stores come from poultry farms and have not been fertilized. Also, eggs not incubated within the proper temperature range for the right time will not develop or hatch. With that question answered, let's move on to raising baby chickens.

If you want to hatch chicks from your eggs, you'll need a rooster. A good rule of thumb is to have 10 to 12 hens per rooster. While it is possible to build an incubator and supervise the development of the eggs, it is easier to let the hens take care of the hatching process.

When a hen prepares to nest, she becomes "broody." This means that she is ready to hatch her eggs. A broody hen will sit on the nest and resist having her eggs collected, whereas a non-broody hen will let you reach under her to collect the eggs. A broody hen may even peck or screech at anyone who comes near. There are ways to discourage broodiness, but it's best to let the hen do all the work of hatching and raising the chicks. Plus, you'll get free chicks!

If you decide to use an incubator, it's recommended to use a forced-air model with an automatic egg-turner, as eggs need to be turned four to five times a day. The temperature inside the incubator should be between 99° and 102°F, while the humidity should remain between 55% and 60%. Chicken eggs will hatch after approximately 21 days. For more information, check with your local cooperative extension service.

Do Chickens Ever Stop Laying Eggs?

Farm chickens have a 4 to 7-year lifespan and can lay eggs for most of this time. They usually lay eggs continuously for about 11 months, after which they stop laying eggs and go through a molting period for several months. In large commercial farms, non-laying hens are sold as feeding them for one to three months is not cost-effective. However, free-range feeding can offset this cost in mini-farms, resulting in lower maintenance expenses during the molting period. Hens may lose feathers and develop bald spots during molting, but this is a normal part of the process. High egg-producing hens tend to molt less and resume laying eggs quicker than lower-producing hens. Molting usually occurs during winter when there is less daylight to stimulate egg-laying. However, egg-laying resumes in the spring as the days become longer.

It is important to note that molting can also be caused by stress factors such as decreased daylight, disease, parasites, overcrowding, insufficient feed or water, and exposure to predators. Therefore, it is crucial to maintain a stress-free environment for the hens.

Raising Chickens For Meat

Raising chickens for meat or eggs can be a fulfilling and educational experience for the whole family. Raising a meat chicken to maturity usually takes 8-12 weeks, while a laying chicken will take about six months to mature. The freshest and most delicious meat can be obtained from matured meat tender and juicy chickens. Therefore, meat chickens are the way to go if you want the best-tasting meat.

Raising broilers can be an affordable way to start. Chicks are easy to acquire and grow fast. Depending on the breed and weight you want them to have at processing, they can be processed and put in the freezer or sold in just 6 to 12 weeks. However, raising healthy livestock requires attention to detail and planning. This applies to broilers, too. To ensure success with the birds you raise, follow basic guidelines on raising baby chicks and growing chickens. Raising chickens for meat or eggs can be a fulfilling and educational experience for the whole family. Raising a meat chicken to maturity usually takes 8-12 weeks, while a laying chicken will take about six months to mature. The freshest and most delicious meat can be obtained from matured meat and tender and juicy chickens. Therefore, meat chickens are the way to go if you want the best-tasting meat.

Calculate the cost

Raising a small flock of chickens can provide enough meat for your household needs. However, if you plan to sell any birds, it's essential to consider the current market prices for each class of meat birds. It's best to avoid competing with retail sales during special promotions. Remember that many customers prefer heavier fryer-type chickens over the lighter range typically found in stores.

To help you with your calculations, consider the following:

- Calculate your production costs and compare them to retail market prices.

- Add a few extra to account for potential losses when purchasing chicks.

- Commercial strains will require about five pounds of feed to age six weeks and eight to nine pounds to age eight weeks.

- Roasters and capons require more feed per pound of meat produced than fryers.

- Keep birds within the time required to reach the desired weight.

- Consider equipment costs, which will depreciate over ten years, and housing costs, which will depreciate over 20 years.

- Estimate litter, heat for brooding, lights, and miscellaneous costs. Also, allow for any payments made for labor for caring for birds, cleaning the house, etc.

- To determine the cost per pound, divide the total cost per bird by the expected market weight. Remember that the ready-to-cook weight will be 70 to 75 percent of the live bird weight.

Meat chicken breeds

Cornish, Plymouth Rock, and New Hampshire breeds are considered the most cost-effective strains for producing meat. These crosses have a rapid feathering and early maturation process and are known for their efficient feed conversion into poultry meat.

Some flock owners prefer using White or Barred Plymouth Rocks, Rhode Island Reds, and New Hampshires for meat production. However, these breeds grow slower than the crosses and require more feed per pound of weight gained. USDA - National Agricultural Statistics Service - Surveys - 2011 ARMS - Broiler Industry Highlights. https://www.nass.usda.gov/Surveys/Guide_to_NASS_Surveys/Ag_R esource_Management/ARMS_Broiler_Factsheet/It is not profitable to raise Leghorn males as meat birds, even if you receive day-old chicks.

Several commercial strains of chicken meat birds are raised for the same purpose.

Broilers or fryers are birds usually slaughtered at 7 to 9 weeks of age when they weigh between 3 to 5 pounds, resulting in a 2 ½ to four-pound carcass.

Cornish game hens are birds that are slaughtered at five weeks of age.

Roasters are birds that are grown out to 12 weeks or longer.

Capons are male birds that are neutered at 3 to 4 weeks and marketed after 18 weeks.

Meat-type chicks are typically purchased on a straight-run basis, which means that both males and females are mixed.

When ordering chicks

Hatcheries located in Minnesota and across the country can be found online. When planning to raise Cornish cross broilers, the most commonly raised breed of chickens, you should aim to have them reach a market carcass weight of four to six pounds within six to eight weeks. However, if you raise other chicken breeds that grow slower, it may take 10 to 12 weeks to reach the same weight. When ordering chicks, you can choose either cockerels (males), pullets (females), or a straight run (mixed batch). Cockerels are a bit more expensive but grow faster than pullets. They may weigh one pound more than pullets when processing at the same age.

It is vital to arrange for processing well in advance. If you plan to raise chickens for your consumption within a town or city, be sure to check the local government ordinances before processing the birds in your backyard, as this is usually not allowed. It would be best to consider vaccinating the chicks against coccidiosis at the hatchery, as it is a cheap way to protect the birds against a common and costly poultry disease. This vaccine can allow you to use non-medicated feed throughout their growth period, which will help keep them healthy.

Raising chickens for meat can be a great idea if you are a small landowner. It has many advantages, such as quick turnaround time, space efficiency, and profitability. You can expect to net 40-50% of your sale price if you can control your losses. In other words, you can double your money in just 10-12 weeks. Moreover, chickens from your pasture or big backyard have superior flavor compared to store-bought chicken.

However, before starting a chicken-raising business, analyzing the market and considering the space requirements is important. It would help if you also decided how you will sell the birds and how many you think you can sell. Keep in mind that chickens only require a little space compared to other livestock, but they still need some room, especially if you plan to run them on pasture. For instance, a flock of 100 meat birds typically roams over about 1/5 acres (6 or 7 thousand square feet).

How will you protect your birds from predators?

When it comes to raising chickens, there are several things to consider. Firstly, where will you source your feed - from a local farmer, mill, or feed store? Additionally, deciding where your birds will be processed would be best. Can you manage the process independently, or is there an abattoir within an hour's drive? Another critical consideration is freezer space - do you have enough to store a batch of birds until they are sold? Finally, it's important to be aware of any regulatory restrictions that may be placed on you. For example, in Ontario, you can only raise and sell up to 300 birds annually, so checking this carefully is important to avoid any issues.

How to Raise Meat Chickens

If you are interested in raising chickens for meat, you're likely to have questions about how to get the best dressed-out bird and how to take care of them. Since the COVID-19 pandemic, individuals and families worldwide have been asking serious questions about becoming self-sufficient in sourcing protein due to shortages in poultry at local grocery stores and supply chain issues. We are here to help our readers by providing information on raising meat chickens, processing chickens, the best chicken breeds for meat, when to process chickens, and more.

1. Choose the Right Chicken Breed for Meat

All breeds of chickens can be raised for meat, but some chickens provide more meat and better flavor than others. When choosing which chicken breed to raise, consider factors such as the chicken's size and the meat's yield. The Cornish Cross and Ranger chicken breeds are the standard meat chicken choices due to their fast-growing abilities and size when dressed.

2. Humane Considerations

There is controversy over the genetics and humane treatment of the Cornish Cross meat chicken. This breed was developed to grow unnaturally rapidly for the commercial industry, which can cause medical complications such as heart problems and leg issues. To avoid these problems, you can choose alternative birds that grow quickly and to a large size with fewer health concerns. One such bird is the Ranger chicken.

3. Time Constraints

If you are working on a deadline, such as preparing for the holidays or sourcing your protein as soon as possible, you may need a bird that finishes out faster than others. In that case, you'll want to research your hatchery's breeds and choose a bird labeled as a meat chicken, such as the Cornish Cross or Ranger.

4. Taste of the Meat

The taste of meat from typical meat chickens or standard dual-purpose breeds varies slightly. The taste has more to do with how you've fed your chickens than the breed of chicken you are raising. Additionally, the older the chicken, the tougher the meat, and chickens used for egg production and then processed for meat are traditionally used for soups and broths.

5. Appearance of the Finished Product

The appearance of the finished product can vary based on the breed of chicken you choose. Cornish crosses are typically white, large pieces of meat, whereas heritage breeds may have yellow or slate-colored skin and a "bony" appearance. If the appearance is important to you, choose a breed that reflects what you'd prefer to see on your dinner table.

We will focus on how to raise large broiler breeds, like the Cornish Cross, and not heritage breeds.

Brooder Basics For Meat Chickens

If you have experience raising layers, your brooder should also be good enough for meat chicks. However, keep in mind that meat chicks grow quite fast. Some of the larger and more robust birds may make it challenging for others to eat and grow if there is not enough space.

You will need a large tub, tank, or box for your brooder. It is best to use a container that is easy to clean and does not promote bacterial growth. The heating temperature for your chicks will remain consistent, regardless of the breed you are raising. You will start with a temperature of either 90 or 95 degrees Fahrenheit, which you will reduce by 5 degrees each week until you can turn it off.

It is essential to monitor your chicks constantly to ensure they are not too hot or too cold based on their behavior in the brooder. If the chicks are too hot, they will try to avoid the heat source, while cold chicks will huddle up and sit tightly beneath it, which can cause suffocation among the chicks at the bottom of the pile.

If you are familiar with the little red feeders and fonts used for layers, you can use those for the first few days to a week after your chicks have arrived. However, as they grow, bullies may emerge in the pecking order, limiting feed to weaker chickens. Hanging feeders (and plenty of them, depending on the number of meat chickens you will be raising) will be beneficial to prevent overeating and bullying.

Meat chickens tend to be lazy and may prefer to lie in open feeders and eat, causing overeating, food hogging, and bacterial growth in the feed due to droppings. To prevent this, try to keep the feeders off the ground and employ automatic feeders. Ensure that all chickens have access to food. The same goes for waterers; they can become quite dirty if not placed strategically off the bedding. Consider using waterer stands, raised waterers, or automatic cup waterers to keep bacteria at bay.

Meat chickens require bedding material that is absorbent and slip-free. Newspapers, bare plastic, or anything that does not give your chickens traction will only exacerbate leg issues, prevent them from walking, and cause them to become trampled and possibly die. Pine wood shavings are the best option for bedding material (never use cedar shavings, which can be toxic to chickens).

Chickens raised for meat need their bedding material changed frequently due to the amount of feed consumed by the meat chickens. Keep this in mind when you plan and consider the amount of bedding you need and how often you need to change it based on the number of broilers you will be raising and the size of your brooder.

Finally, make sure your meat chickens are kept clean and live in dry conditions. Bacteria love to grow in a wet, dirty brooder.

Feeding Instructions for Meat Chickens

When raising Cornish Cross and some Ranger breeds, it's crucial to feed them correctly based on their age to prevent overeating, poor meat quality, and medical conditions. For the first week, let your broilers have free access to their feed by keeping their feeders full. During this time, chicks require much protein to grow correctly, and free choice is acceptable. Feed your meat chickens a 20% protein chick starter during the first three weeks, then switch them to an 18% protein grower feed. After one week, feed your chickens 12 hours on (free-choice) and 12 hours off. Don't forget to take their feed away during the off hours to prevent overeating.

Type of Feeds for Meat Birds

When you get your chicks, make sure they eat chick starters. It's usually the same for layers as it is for broilers (but always ensure it has at least 20% protein content). You can choose to feed medicated or non-medicated feed. Medicated feed protects chicks from contracting coccidia when they are most vulnerable to intestinal infection. Coccidiosis is a potentially deadly intestinal tract infection caused by protozoa found in chicken droppings. Therefore, medicated feed might be worth it initially. When your chicks reach three weeks of age, switch them to a grower/broiler formula. During this period, provide your meat chickens with grit to aid digestion.

Moving Day

At some point, your chickens will outgrow their brooder. Always ensure that you adjust their living space as they grow to avoid overcrowding. Some sources suggest that Cornish Cross only needs about two square feet of space per adult chicken. But if you can give them more, do so. Smaller spaces invite pecking, pileups of feces, and overweight chickens. More space gives your chickens the freedom to move, flap their wings, and exercise their bodies. It's good to allow them the room to wiggle, move, and act like chickens. We prefer free-ranging our rangers. The Cornish can free-range, but they are less willing to do so in our experience.

In the Coop

Whether you're keeping your meat chickens in a coop or chicken tractor, ensure that you maintain the same mindset regarding feeders and waterers. Keep them clean, full, and off the ground to allow everyone to eat what they need to grow and develop into happy, meaty chickens. Also, ensure that you can keep the feeders and waterers clean to prevent bacteria growth that can cause illnesses in your chickens.

Knowing When to Process Your Chickens

Processing your chickens at the right time is crucial to getting the best out of your birds. If you are raising Cornish Cross, they should be processed at around 8 or 9 weeks. We have a detailed guide available for more information on the matter. If you've followed our feeding schedule, you can expect to have some delectable meat chickens at that time, with very few losses.

However, if you are raising heritage breeds or rangers, your processing date will likely be pushed back by a few weeks to a few months, depending on the breed you have chosen to raise. It's important to understand your breed and always research the recommended processing time frame, as waiting too long to butcher a chicken after it has reached maturity will result in a tough or stringy carcass, meaning that it's best suited for soup.

In conclusion, understanding the breed you are raising and the desired weight for processing is crucial when raising meat chickens. Always research the appropriate processing time frames and take care to ensure that you do not wait too long to process your chickens.

Beekeeping On Your Mini-Farm

Bees have a vital role in our ecosystem, and beekeeping offers several advantages. The primary benefit is the production of honey, which is a valuable food source. Moreover, beeswax is used in various industries, such as cosmetics, polishes, and pharmaceuticals.

Here are ten reasons why bees are essential for the environment and humanity:

1. Pollination: Bees are excellent pollinators, helping to produce food in the form of fruits, nuts, berries, and seeds. This includes leaf and root crops, such as peas, beans, and kale.

2. Food production: Bees are responsible for cross-pollinating many of the foods we eat. Without bees, we would not have delicious treats like strawberry ice cream, apple pie, and blueberry muffins.

3. Nutrient value: Bees improve the nutrient value of the crops they pollinate, which can have a direct impact on human health.

4. Seed production: Bees help ensure that seeds are set for many plants, allowing farmers to sow crops for the following year.

5. Global economic benefit: Bees are hugely important to us, as they pollinate agricultural plants. Their global economic benefit was estimated to be EUR 265 billion in 2015.

6. Honey production: Honey is the most popular product that comes from bees. Bees collect nectar from flowering plants or honeydew from coniferous trees and store it in their honey stomach. The harvest is stored in the hive, processed and dried by the bees, and then harvested by beekeepers.

7. Harvesting honey: Beekeepers use honey extractors to remove honey from the honeycombs. The honey is then filtered and put into jars, where it is left to settle for a few days.

8. Wax production: Beeswax is used in many industries, including cosmetics and pharmaceuticals.

9. Propolis: Bees use propolis, or bee glue, to protect their hive from bacteria and viruses. It has antiviral, anti-inflammatory, and antibacterial properties.

10. Biodiversity: Bees help maintain biodiversity in our ecosystem, which is essential for the health of our planet and our survival.

The bee colony is a source of many valuable products with healing properties. One such product is wax, which is primarily used in the cosmetic and pharmaceutical industries. Bees secrete the wax from their glands to construct the honeycombs. The beekeeper removes the old, brown honeycomb, melts it down, and removes impurities, resulting in light and pure wax. This wax is then cast into new wax center walls and returned to the colonies or recycled in other ways. Beeswax is a valuable raw material used to make beeswax candles, among other things. It takes the wax from an entire beehive to make just one candle. Beeswax candles emit a soft light and a pleasant aroma, making them a popular choice.

Bees produce various valuable products, including pollen, bee glue (also known as propolis), and royal jelly. Pollen is collected by bees from flowers, mixed with nectar, and taken back to the hive. It is commonly used as a nutritional supplement because it is rich in vitamins, minerals, and proteins, which can improve mental capacity and strengthen the immune system. Beekeepers can collect between 30-60 kg of pollen per year by attaching a pollen trap to the hive's entrance.

Bee glue, or propolis, is made by bees using resin from different trees. They use this glue to protect themselves from bacteria and fungi. A single colony can collect up to 500 grams of resinous mass annually. Propolis is a natural antibiotic with anti-inflammatory effects and is used in many medicinal products. It is available in capsule form, medicinal ointments, creams, or herbal drinks.

Royal jelly is the most valuable bee product, which is produced by young worker bees using honey, pollen, and glandular secretions. The queen bee is the only bee that eats royal jelly, which can extend her life expectancy to 50 times longer than her colony. The jelly is used in the cosmetics and medical industries, where it is recognized for its revitalizing effects and as a natural remedy for viral infections.

Bees are essential for maintaining human and planetary health. They play a crucial role as pollinators and produce various products with medicinal properties. While bees are managed by humans, there are over 20,000 known bee species worldwide, with over 4,000 native to the United States, most of which are wild. The importance of bees in promoting food security and biodiversity is widely recognized. However, factors such as pesticide use and urbanization have led to a decline in bee populations, negatively affecting many of the Earth's ecosystems. The loss of bees would not only affect honey supplies but also have significant impacts on world food security and biodiversity. Therefore, it is crucial to protect and preserve bee populations.

Why are bees important?

Bees are important for several reasons. They play a significant role in maintaining healthy ecosystems, have historical importance, and contribute to human health.

Health benefits:

Honey is one of the main reasons people value bees, but not all bees produce honey. Honey is a natural sweetener and has many potential health benefits. For thousands of years, people have used bees and bee-related products for medicinal purposes. Some researchers have noted claims that it has antioxidant, antimicrobial, anti-inflammatory, and anticancer properties.

Traditional medicine uses honey to treat various conditions such as eye diseases, bronchial asthma, throat infections, tuberculosis, thirst, hiccups, fatigue, dizziness, hepatitis, constipation, worm infestation, hemorrhoids, eczema, ulcers, and wounds. Beeswax, another vital bee product, has previously been used in waterproofing and fuel. It currently has benefits for health and features in many skincare products. Additionally, pharmaceutical industries use it in ointments.

Other bee products that can benefit human health include propolis, bee bread, bee pollen, royal jelly, beeswax, and bee venom. In a 2020 study, scientists found evidence that melittin, a component in honeybee venom, could kill cancer cells.

Pollination:

Bees play a crucial role in pollination, using the hairs on their bodies to carry large grains of pollen between plants. Around 75% of crops produce better yields if animals help them pollinate. Of all animals, bees are the most dominant pollinators of wild and crop plants. They visit over 90% of the world's top 107 crops. In other words, bees are essential for the growth of many plants, including food crops.

Historical importance:

People have been working with bees around the world for millennia. The significance comes from the direct harvesting of honey and beeswax and cultural beliefs. For example, the Ancient Greeks considered bees to be a symbol of immortality. Meanwhile, native northern Australians used beeswax when producing rock art. For history experts, bee products are a crucial aspect of archaeology. This is because beeswax produces a "chemical fingerprint" that people can assess to identify components in organic residue.

Society and the environment:

Bees are intelligent creatures, and people have applied knowledge of their mannerisms and social interactions when creating human initiatives. For example, researchers have suggested that studying the actions of bees could help experts develop emergency plans to evacuate people from an overcrowded environment. Observing honeybee dances can also help scientists understand where environmental changes occur.

How does this affect humans?

Bee numbers are declining due to farming practices, global warming, and disease, among other reasons. Experts are concerned about the impact on world food supplies, especially fruits, nuts, and vegetables. They say that without bees, there will be no more nuts, coffee, cocoa, tomatoes, apples, or almonds, to name a few crops. This could lead to nutritional deficiencies in the human diet, as these products are essential sources of vital nutrients.

Additionally, the emerging medicinal properties of bee venom and other bee products may never be accessible without bees to provide them. In financial terms, the pollination of fruits and vegetables by wild bees across the United States has a high economic value. One 2020 study found that wild bees were responsible for a significant portion of net income from blueberries. There is a direct link between farmers' economic yield and bees' presence. In 2012, experts estimated total pollination to be worth $34 billion, with a large portion due to bees.

How can you assist bees? Green gardens and backyards can serve as crucial resources for these insects. By growing native flowers and allowing weeds to thrive, you can help improve the health and numbers of bees by providing them with food and shelter. Limiting landscaping activities like mowing or pruning can also benefit bees by increasing the amount of vegetation available.

In addition to benefiting bees, a 2019 study revealed that increasing rural areas within urban settings can enhance human mental and emotional well-being.

Citizen science initiatives allow nonscientists and volunteers to contribute to research by reporting what they see in their local area. Experts can use this information to better understand what is happening in a particular area or country. For instance, a 2020 citizen-based study found that squash bees are widely spread and prefer farms with less soil disturbance.

Similarly, the 2007 Great Pollinator Project, a partnership in New York, encouraged the public to observe bees and record the types of wildflowers they visited. Such findings help scientists find effective ways to protect bees. However, this depends on people correctly identifying the species. Therefore, learning about bee species and habits can help individuals protect them.

Preserving Your Produce

· Utilize Your Crisper.

· Wrap Your Greens.

· Keep Some in The Dark.

· Pick The Right Time To Wash.

· Treat Them Like Flowers.

· Make Herb Cubes.

Tips On How To Store And Preserve Your Garden Harvest

These tips help you get the most out of every harvest so nothing goes to waste!

It's not uncommon for many of us to pick our homegrown fruits and vegetables and toss them into the no-man's-land of the fridge, only to later find they wilted, got mealy, or worse, became a complete science project.

Throwing away what you grew is like throwing away your hard work, time, and money. So what can you do? How you handle and store your produce will determine how long of a shelf life it will have, putting more of it on your table and less into the trash.

If you want your garden harvest to last longer, here are some tips for storing it properly.

1. Bring it indoors

As soon as you've harvested your produce, bring it inside instead of leaving it outside. Sunlight can over-ripen and soften the produce. However, if you've picked anything too soon, you can place it in a sunny spot, such as a kitchen window, deck, or patio, to finish the ripening process.

2. Chill or keep at room temperature

Some fruits and vegetables benefit from refrigeration, while others should be kept at room temperature. Keep vegetables such as asparagus, summer squash, yellow squash, green beans, broccoli, cauliflower, Brussels sprouts, peppers, green beans, berries, and cucumbers in the fridge, ideally in your crisper drawer. Store mushrooms in paper bags to prevent them from getting mushy or moldy, and keep ears of corn in their husks. Lettuce and other leafy greens should be stored in airtight bags. You can also place asparagus stalks in a glass of water in the fridge. Most fruits, such as melons and citrus, should be stored on the counter, along with tomatoes that are out of direct sunlight. Stone fruits like apricots, peaches, plums, and nectarines can be kept on the counter until they ripen, then placed in the fridge to prolong their shelf life. Cherries, however, should be immediately placed in the fridge in an airtight container or bag.

3. Use the crisper

The crisper drawer was designed to help decrease water loss (transpiration) in fruits and vegetables, making it very useful in making them last. You can adjust the humidity vent to control the airflow in the drawer.

4. Wrap your greens

Leafy greens are best enjoyed immediately after harvesting when they are crisp. They are often vulnerable to moisture loss and wilt quickly. However, you can wrap them with a damp paper towel and store them in a sealed, airtight bag to maintain humidity if you can't eat them immediately.

5. Keep some in the dark

Onions, garlic, shallots, winter squash, and pumpkins prefer a drier climate with average humidity and will last longer when kept in a dark, cool cupboard.

6. Wash at the right time

It's generally best to wait until you're ready to eat your produce before washing it, as washing it prematurely can significantly reduce its lifespan.

7. Handle with care

Fruits and vegetables can be very delicate, and any injury, from a bruise to a scratch, can cause them to ripen or rot more quickly. If one piece is damaged, such as an eggplant, eat it first. Remove any rotten produce from the group, as one rotten item will spoil the others more quickly.

8. Treat them like flowers

Harvested herbs such as mint, basil, and cilantro can be kept in a glass of water in the fridge, like a bouquet. Place a plastic bag over the top to help balance the moisture. Refrigerate most herb "bouquets," but store basil on the kitchen counter where it can get some sunlight. Alternatively, you can loosely wrap your hard, woody herbs, such as sage, rosemary, thyme, and oregano, in a damp paper towel and place them in a sealed bag. The paper towel will keep them moist enough to prevent them from drying.

9. Make herb cubes

You can save your herbs by making frozen herb cubes. Break up herbs by hand and add them to ice cube trays. Fill the cubes with olive oil or coconut oil and place them in the freezer to solidify. Transfer to a labeled freezer-friendly container. When you're ready to use them, pop them into your favorite dishes.

10. Freeze your surplus

Freezing your harvest is an excellent way to take advantage of peak-season produce all year. First, wash items and dry them thoroughly, as moisture can cause rot. Most fruits and vegetables freeze better raw, but some are best blanched beforehand. Blanching stops enzyme actions that can cause loss of flavor, color, and texture and helps preserve nutrients.

11. Make green cubes

Add a bunch of spinach or any leafy green that you can't use immediately to a blender with enough water to process. Freeze the mixture into ice cubes and store them in a sealed container. Green cubes are a way to add extra nutrients to your smoothies, soups, and casseroles.

12. Old school canning

Although canning requires some equipment (canner, canning jars, lids and bands, and a jar lifter), it is an excellent way to preserve the nutrients and flavor of vegetables like asparagus, beans, carrots, corn, peas, and mixed vegetables. For a detailed explanation of the canning process, go to the National Center for Home Food Preservation website.

Selling Your Produce

If you're looking to sell your produce, there are three main options available to you: dealing directly with restaurants, selling through farmers' markets, or creating your own market. You can also reach out to small stores or work with local restaurants. In addition, certifications and pricing are important factors to consider, as well as the support of local consumers.

Create Your Own Market

One of the most common ways of creating your market is through a Community Supported Agriculture (CSA) program. With a CSA, individuals directly support a local grower by purchasing a "share" of the harvest at the beginning of the growing season. In exchange for this investment, they receive a weekly supply of everything that is harvested. While CSAs are typically done on a reasonably large scale, there's no reason why you couldn't create your own version. You could survey friends and neighbors to get a sense of what people would be most likely to want and balance that with what you're most likely to be able to produce. You don't necessarily need to have people buy in before you start growing, especially as you're starting up. You may be more comfortable selling on an as-available basis and not being indebted before you're confident of your harvests. Many CSAs deliver, while most have a set pickup time and place. If you can provide fresh produce every Wednesday for pickup on the way home, you can develop a regular clientele. Alternatively, to ensure regular sales, you can make deliveries and collect money monthly (presuming that folks will only sometimes be home when you drop off).

In addition to the above, you can consider selling to retirement communities, apartment complexes, and other centralized communities. Retirement complexes, in particular, are great options as those folks often need easy access to groceries and fresh, homegrown produce. Many older people miss the days of growing their food, and this is the last generation who knows where food comes from and tends to have significant climate- and place-specific advice.

If you are looking to sell your harvest locally, there are a few ways to go about it. One option is to work with local restaurants by talking to the owners and chefs. It's best to start with the chefs and find those who are willing to cook with seasonal produce. Selling to restaurants can be a great way to build relationships and support your local food economy. However, you need to be consistent in delivering the right quantity and quality at the agreed time. Even the most dedicated chef can only be so flexible, and running out of a critical ingredient mid-shift can cause problems.

Another option is to sell at farmers' markets. This gives you more flexibility than selling to restaurants, as you have your own or shared market stall. If, for any reason, you can't deliver the expected harvest, the repercussions are less severe. You can start by exploring your local farmers' markets for ideas and advice and consider partnering with someone who sells complementary produce. Sharing a stall or selling to a grower who's already there can also be a good option.

Lastly, you can reach out to small stores such as corner shops and independent markets. This approach gives you the opportunity to sell your produce over the course of a week instead of just a day at the farmers' market. Let the store owners set the prices and percentage they'll take. If you're not happy with their terms, suggest an alternative, but be prepared to look elsewhere if necessary.

Certifications and Pricing

Regarding your other two inquiries, Obtaining organic certification can be quite expensive for small-scale growers, but the costs are gradually decreasing. ATTRA and the USDA have information on how to obtain certification. Unfortunately, as industrial agriculture moves into the organic market, the meaning of the term "organic" is becoming less and less meaningful. Whether you choose to pursue organic certification or not, it's important to let your buyers know that you use sustainable and organic growing methods. Pricing for your produce can vary greatly and is constantly changing. I suggest looking at market prices (such as those at Whole Foods or local co-ops) or asking owners to determine a fair price.

Consumers: Support Local Growers

Living in a consumer-driven society gives you a certain amount of influence. Use that power to support local growers. Find them through Local Harvest and ask the owners and chefs of your favorite places if they use local sources. Make it easy for others to find local suppliers by providing them with information from Local Harvest or ATTRA. Localized economies are the foundation for all things sustainable, and eating locally-grown food is the best choice for yourself and your family.

How to Make Money With Extra Garden Produce

Join a Small Farmers' Market

If you have surplus produce and are willing to commit to an entire season, selling at a farmers' market is a traditional way to make money. I recommend researching local options to find a market that makes sense for your growing scale. Generally, larger markets are more difficult to get into and may require a commitment to selling twice per week throughout the season or a spot on a standby basis. In contrast, smaller markets are more lenient and let you pay for a table on the days you attend. However, these markets get much less foot traffic, and vendors tend to set lower prices.

If you have extra garden produce and you're looking for ways to sell or give it away, there are several strategies you can use. One option is to set up a farm stand in your front yard. This can be as simple as a cooler near the end of your driveway or a sturdy wooden structure with a mini-fridge. Keep in mind that some crops are better suited to farm stand conditions than others, and you'll need to check local regulations to make sure it's legal. If you prefer not to have a farm stand, you can team up with another local farm stand to sell your surplus.

Another strategy is to use social media to connect with people who are interested in buying fresh produce. You can list your available crops and prices on crop swap sites like BigBarn and RipeNearMe or post about your surplus on social networking sites like Nextdoor. You can also sell your produce through ads on Freecycle or Craigslist.

If you have excess produce that you can't sell, you can turn it into value-added products like jams, pickles, and other preserved foods. However, be aware that regulations on canned goods can be strict, especially for items with low sugar or salt content. Another option is to bundle fresh vegetables with recipe cards for easy meal ideas.

If you have an abundance of herbs, you can dry them and make homemade herbal teas and seasoning blends. Dried herbs are popular at farmers' markets and have a long shelf life, making them easy to store and ship.

If you don't want to sell your excess produce, you can give it away. You can put a "free" sign in your front yard, give veggies to visitors in used plastic bags, or cook a meal for a neighbor. You can also donate to organizations like Ample Harvest or the Food Rescue Locator, which help connect gardeners with nearby food pantries.

If you have specialty vegetables, you can sell them to local restaurants. Many chefs are interested in featuring locally grown produce on their menus, even in small quantities. To increase your chances of selling, grow unique or high-quality crops, and contact restaurants early in the growing season.

Finally, if you can't sell or give away your excess produce, you can compost it. Compost is a nutrient-rich garden amendment that can improve your soil. Make sure to add a mix of "green" and "brown" materials to your compost pile for the best results.

To succeed in selling your garden produce, keep it simple and grow no more than five varieties at first. When harvesting crops like zucchini, beans, and cucumbers, it's best to pick them when they're small for extra-tender produce.

It's crucial to think through your storage and distribution strategy, especially for high-value crops like lettuce and fresh berries that only look good for a day or two after harvesting. Do you need to invest in a second refrigerator or another form of cold storage to keep them fresh for longer?

Sell only the best produce that is perfect. Though it may be tempting to keep the choicest vegetables for yourself, selling or giving away only the best produce will earn you a good reputation. If you sell inferior garden goods, you might never bounce back.

Presentation is everything. Think about how you want to display your produce at the farmers' market or your farm stand. Wicker baskets and nice tablecloths can help put your garden bounty in the best light to encourage sales.

Pay attention to the niche you can fill. Growing something unique that no one else has is the best way to make money from the garden. Do some local research to learn what gap you can fill. Is there a market for dried Indian corn or winter squash? Are people clamoring for vegetable starts in the spring? Plan your growing strategy accordingly.

Consider establishing a U-pick option. Many people are willing to pay a premium to harvest their vegetables. Think of the appeal of u-pick apple orchards—people love putting in a bit of labor to feel extra connected to their food. If you're comfortable letting customers into your garden, let them pick what they want and pay by the pound at the end.

There's no reason to feel overwhelmed by your garden's success. Whether you're looking to make some money or merely to bless others with your bounty, there are plenty of ways to take care of any garden surplus without waste. Be intentional about your extra produce plan now, and you'll be making giant strides to help us all enjoy a more localized, sustainable food system.